Science vs Religion?

Science vs Religion?

Intelligent Design and the Problem of Evolution

STEVE FULLER

polity

First published in 2007 by Polity Press
Reprinted in 2008

Polity Press
65 Bridge Street
Cambridge CB2 1UR, UK

Polity Press
350 Main Street
Malden, MA 02148, USA

ISBN-13: 978-07456-4121-8
ISBN-13: 978-07456-4122-5 (pb)

A catalogue record for this book is available from the British Library.

Typeset in 11 on 13 pt Bembo
by SNP Best-set Typesetter Ltd., Hong Kong
Printed and bound in the United States by Odyssey Press Inc.,
Gonic, New Hampshire

For further information on Polity, visit our website: www.polity.co.uk

Contents

Introduction

This book deals with the philosophical and sociological dimensions of the signature intellectual struggle of the modern era: *science vs religion*. However, the struggle does not have much intellectual depth when posed in such terms. Of course, the feeling that science and religion pull in opposing directions has been integral to the experience of "being modern." As sources of authority in the wider society, science and religion have been often polarized – but in ways that do not correspond very clearly to substantive intellectual differences. From the standpoint of the sociology of knowledge, this is not surprising: Abstract arguments of principle call forth rival rhetorics designed to include and exclude certain groups based on views the parties are suspected to hold about matters *other* than those formally contested.

A good case in point is the now 150-year dispute between evolutionists and creationists for scientific authority – especially over the nature of life. At first, this dispute was confined to the anglophone world but now has acquired global dimensions. In the United States, where the conflict has always been most heated, it amounts to high-minded shadowboxing vis-à-vis deeper political and economic struggles that are most openly expressed on the floor of Congress or in an election campaign. In this context, creationists appear as defenders of traditional moral values that are under threat by evolutionists, whose liberalism promises a more open sense of what is right and wrong. However, as should become clear in these pages, these pervasive stereotypes effectively interfere with the deeper intellectual issues at stake in the dispute.

Not surprisingly, then, readers will wonder where I stand on the substance of evolution–creation dispute. I believe that the version of creationism nowadays called "intelligent design theory" (or IDT), which takes inspiration from the Bible but conducts its business in the currency of science, was responsible for the modern scientific world-view that evolution nowadays exemplifies so well. Even those who were led to reject IDT, not least Charles Darwin, began by assuming its vision of nature as a rational unity designed for human comprehension. In contrast, the general evolutionary perspective that Darwin ultimately championed has many cross-cultural precedents but these have tended to discourage systematic scientific inquiry, stressing instead the need to cope with our transient material condition in an ultimately pointless reality (Fuller 2006b: ch. 11). I believe that to lose touch with the creationist backstory to modern science would be to undermine the strongest reason for pursuing science as a transgenerational universalist project that aims to raise humans above the animals.

In short, contrary to what advocates on both sides of this dispute appear to believe, IDT provides a surer path to a "progressive" attitude to science than modern evolutionary theory. Darwin's theory of evolution by natural selection managed to create such a furore in the West – but not in the East – because his careful organization of the scientific evidence appeared to imply that the pursuit of science itself is ultimately meaningless: the diversity of life would seem to lack the cosmic design that had inspired previous generations of Christians, Jews, and Muslims to study nature systematically. In effect, Darwin undermined what had always been a fundamentally religious motivation for doing science: the ennoblement of humanity, the species created in God's image.

If you find my position disorienting, keep in mind that nothing I say in these pages speaks against the *empirical* success of modern evolutionary theory – the only question is whether it should be taught as the single paradigm concerning the nature of life.

Whether science's historic religious motivation – especially the idea of a unified and rational conception of reality – can be sustained in a culture completely bereft of monotheism remains an open question. Interestingly, while there is a plethora of evolutionary accounts of religion, hardly any exist of science itself. Instead, one is treated to a battery of weak arguments, collectively known as "evolutionary epistemology," that are really addressed to the evolution of a more primitive concept, *knowledge*. Thus, science is portrayed as a very elaborate extension of, say, basic survival skills, animal curiosity, economized effort, innate bias, aesthetic preference, or, for those who like their Darwinism mixed with cultural relativism, learned adaptive advantage. Of course, it would be well within the naturalistic spirit of evolutionary theory to conclude that, from a strictly biological standpoint, there is no selective advantage for humans to pursue science indefinitely. For example, science may be ultimately responsible for too many people consuming too much energy to make for a sustainable ecosystem.

While no one seriously doubts evolutionary theory's tremendous impact on our understanding of natural and social phenomena, there remains a problem to which philosophers have been especially sensitive: how exactly one defines this "theory" which has had so much impact. The physical sciences offer a clear sense of how theories are constructed and hypotheses specified, tested, corrected, and rejected. This is largely because the history of physics and chemistry provides a track record of at least three centuries in which these philosophically relevant stages have been identified and analyzed. It is not by accident that philosophers of science as recent as Thomas Kuhn (1970) have based their own accounts exclusively on these disciplines.

In contrast, and contrary to how the matter is sometimes portrayed in folk histories of science, evolution became the common theoretical paradigm of the life sciences only once natural history and experimental genetics were unified in carefully crafted textbooks and popular histories as the "neo-Darwinian synthesis." This feat, a product of the 1930s and 1940s, was largely the work of two of the most intellectually ambitious and religiously inspired biologists of the 20th century, Theodosius Dobzhansky and Julian Huxley, who figure periodically in the pages of this book. If it appears that the life sciences have been brought closer together in the aftermath of their efforts, that is due at least as much to the concentration of financial backing and political interest as to genuine intellectual breakthroughs in molecular biology (Gilbert 1991; Kay 1993).

Nevertheless, the various branches of biology nominally unified by the synthesis have mostly tried to pursue their own research trajectories. As a result, each biological speciality has a somewhat different understanding of how the neo-Darwinian synthesis is supposed to work, typically biased to the author's own speciality (cf. Segerstrale 2000). Thus, only the death of one of the parties could end a highly public quarter-century spat over evolution's implications between the paleontologist Stephen Jay Gould, from the natural history side, and the zoologist Richard Dawkins, from the genetics side, of the synthesis. Their numerous disputes (neatly organized in Sterelny 2001) very much resemble the ones to which social scientists have grown accustomed over the last century or more, whereby the "softer" more history-oriented side of the field (cf. Gould) argues for multiple causation in local settings, while the "harder" more laboratory-oriented side (cf. Dawkins) appeals to universal laws applicable in all settings. So, then, what is the epistemological status of natural selection: one of many factors operative in the evolution of life – and perhaps not the most important – or an inescapable cosmic principle?

As the neo-Darwinian synthesis made its way into biology textbooks in the 1950s, philosophers tried to confer on the theory some of the logical rigor associated with classical and post-classical physics, where it was possible to state fundamental laws, ideally expressed in mathematical terms, from which hypotheses could be deduced given certain auxiliary background assumptions, and then predictions tested under specific conditions. With that in mind, Antony Flew refined Julian Huxley's original formulation of the synthesis to propose the following as the basic deductive structure of modern evolutionary theory (Oldroyd 1980: 118–19):

GRI + LR → SE
SE + V → NS
NS + T → BI,

where GRI = Geometrical Ratio of Increase; LR = Limited Resources; SE = Struggle for Existence; V = Variation; NS = Natural Selection; T = Time; BI = Biological Improvement. While this set of chained formulae certainly captures the spirit of Darwin's theory, there is considerable scope for interpretation as to the range of values each of the variables may take. For example, the basic *unit* of selection is left wide open: does GRI refer to a population of individuals of the same species, individuals of different species living symbiotically in a common environment, or perhaps some reproductively relevant part of an individual organism, such as a gene or some still smaller unit? The presence of these ambiguities reflects just how *little* that original theory of Darwin's anticipated of modern developments in the life sciences.

Combined with Darwinism's undoubted cultural resonance, many philosophers and social scientists, even those who style themselves as "evolutionists," have been loath to treat the theory as more than an extended analogy, or multipurpose model. They have resisted the temptation to explain everything as a direct or even indirect result of biological survival. Examples include the philosophers Karl Popper (1972) and Stephen Toulmin (1972), the psychologist Donald Campbell (1988), and the sociologist Walter Garrison Runciman (1983–97). Of course, some like philosopher Daniel Dennett (1995) and socio-biologist E. O. Wilson (1998) quite happily grant the neo-Darwinian synthesis the scientific authority formerly accorded to Newtonian mechanics, but they remain in a minority. Nevertheless, generally speaking, as the philosophical paradigm of scientific theorizing has shifted from Newton to Darwin, the standards for what counts as a well-formed scientific theory have likewise changed. In effect, the standards have been loosened, so that a good scientific theory may model many domains of phenomena adequately without necessarily bringing them all together under one unified explanatory structure.

These looser standards bear on the difficulties involved in offering a fair assessment of the merits of intelligent design vis-à-vis evolution as theoretical accounts for the history of life of Earth. The difficulties are epitomized by the ubiquitous demands for "evidence" for this or that claim that one or the other side has made. While, in principle, the request for evidence is always to be welcomed in scientific inquiry, it can only have forensic value if the two positions are agreed that the requested evidence would in fact differentiate them. This is easier said than done. It would require that "intelligent design" and " evolution" be discussed at the same level of generality – which they tend *not* to be.

On the one hand, the phrase "intelligent design" is usually restricted to a group of mostly American scientists and ideological fellow-travelers who over the past quarter-century have been variously trying to shore up the scientific credentials of creationism. US courts, when given the chance, have ruled creationism to be religion and *therefore*, according to the prevailing idiosyncratic reading of the US Constitution, not science. Intelligent design is rarely treated – even by its adherents – as the venerable research tradition that it is, which reaches back to scientists who are regarded as contributors of lasting significance, even by evolutionists. I mean here not only the physical scientists associated with the Scientific Revolution of the 17th century, like Isaac Newton, but also the inventor of modern biological nomenclature, Carolus Linnaeus, and the discoverer of the mathematical laws of heredity, Gregor Mendel, both of whom were by today's standards "special creationists," i.e. believers that each species was created *de novo* by God.

On the other hand, proponents of evolution generally embrace a broad church or "big tent" approach to their theory when dealing with intelligent design theorists. Thus, historically significant disagreements – over, say, the exact role of natural selection or its degree of "blindness" – are publicly suppressed in favor of an abstract definition of evolution as " descent with modification," words from Darwin that, like so much else he said, no longer mean quite what they did in 1859. But exactly what they mean now is left in a tactful limbo. This point perhaps came out most clearly in June 2006 when 67 national academies of science declared themselves in support of evolution in its worldwide struggle against "neo-creationism." However, the fine print revealed that consensus had been secured only on matters like the age of the Earth but not on the means by which life evolved.

Given this rhetorical mismatch, evolutionists can absorb virtually all criticisms made by intelligent design theorists by shifting between different models of the evolutionary process, while intelligent design theorists look like mere upstarts carping from the sidelines, lacking an alternative scientific base of their own. Intelligent design theorists are thus treated as if they do not have the right to reinterpret the extant evidence for evolution to support their own case, even though much of that evidence and its conceptualization came from people who either did not endorse evolution or did not feel compelled to decide between evolution and creation.

The fault here lies on both sides, but in particular intelligent design theorists appear to suffer from the misapprehension, associated with Kuhn (1970), that practitioners of the dominant scientific paradigm enjoy a monopoly over the interpretation of the science's history. I have strongly critiqued this point of view (Fuller 2000b). In this context, a big problem with toeing the Kuhnian line is that intelligent design theorists are then forced to bear an insurmountable burden of proof, namely, to arrive at a research program that is radically distinct from what has so far transpired in the history of biology. This is neither possible nor should it be necessary. Indeed, intelligent design theorists would do well to reclaim the likes of Newton, Linnaeus, and Mendel as their own. A useful role that a critical work of philosophy or sociology can perform is the Rousseauian one of enabling people to realize that restrictions on their thought and action have been largely self-imposed. If this book helps to balance the ledger between evolution and intelligent design, then it will have served its purpose.

Chapter 1 continues to elaborate the points raised initially in the Introduction, namely that the supposedly ongoing conflict between science and religion masks deeper concerns that cut across the science–religion divide but which go to the heart of "Western culture," understood broadly. Here I argue that church-based restrictions on scientific recognition in higher education have been historically the source of more conflict than the actual content of the views held by people on the fair and foul sides of such restrictions. Once the church no longer controlled the universities, these conflicts tended to subside. However, the spirit of Darwin's theory of evolution poses a deeper challenge to the creationist assumptions of Western science, which tend to be taken for granted, and hence forgotten, today. The chapter concludes by showing that proponents of evolution display internal intellectual divisions to those of creation that may in fact run deeper than the differences between evolution and creation themselves.

Chapter 2 addresses the roots of contemporary intelligent design theory, the content of which stresses the Newton-inspired nonconformist elements of scientific creationism, specifically the prospective merger of divine and human intellects. One strand of this trajectory moves through the history of positivism in social science. However, the fascination of today's leading funders of intelligent design, Seattle's Discovery Institute, with artificial intelligence points to an affinity with "transhumanist" thinking that also reaches into the more theistic aspects of, say, Julian Huxley's version of evolution, which are concerned with species improvement, perhaps to the point of immortality. This chapter draws out some of the implications for the organization of the sciences if one takes seriously the intelligent design idea that biology is ultimately "divine technology" that may be re-engineered to purpose.

Chapters 3 and 4 present contexts in which the intellectual differences between evolution and intelligent design appear, respectively, narrowed and widened. Chapter 3 focuses on the centrality of "complexity" in both creationist and evolutionary accounts of the nature and content of the life sciences. Here I conclude that the topic is likely to reduce the differences between the two positions to alternative interpretations of computer simulations. In contrast, chapter 4 treats the two positions in the US legal arena, where "science" and "religion" are constitutionally interpreted as mutually exclusive terms. This results in enormous distortions of the history and philosophy of science, not least in terms of the respective significance of "naturalism" and "supernaturalism" as metaphysical frameworks for the development of science.

Chapter 5 considers the 21st-century prospects for moving away from Darwinian evolution and toward a renewed sense of intelligent design. I begin by comparing the fates of two kindred spirits of the 19th century, Marx and Darwin, both of whom put forward "historicist" theories that Karl Popper equally faulted on scientific grounds. Yet Darwin succeeded where Marx failed in the 20th century. To be sure, Darwin's achievement has been largely rhetorical, as the theory of evolution by natural selection loosely constrains a vast range of biological disciplines, more in the spirit of a political party platform than a mathematical physical theory. I show that this looseness enables modern evolutionary theory to appear much more unified than the comparably arrayed disciplines in the social sciences, without having to encompass the social sciences into a kind of "sociobiology." After considering the well-avoided historical precedents for such a massive synthesis, I turn to its renewed prospects in a secular version of intelligent design, inspired by Julian Huxley, called "transhumanism." However attractive such prospects may be, they need to be tempered against possible nativist and racist backlashes inspired by an updated understanding of Darwin.

The book's concluding reflections bring together the principal themes, placing them in the context of both larger philosophical lessons (which stress the differences *within* – not *between* – science and religion), and more strategic considerations of pursuing the debate between evolution and intelligent design as an ongoing cultural conflict in classrooms *not only* in the United States.

My interest in the topics covered in this book reach back to my Jesuit training at Regis High School, New York, which unwittingly (so I guess) left me with the sort of intellectual sophistication that confounds clear spiritual commitment. My appreciation of the intertwined, if not indistinguishable, character of science and religion was deepened by my training in history and philosophy of science, first under Mary Hesse at Cambridge and then under J. E. McGuire at Pittsburgh. While Professor of Sociology at the University of Durham, I developed an undergraduate sociology of science course that stressed the Judeo-Christian and Muslim religious origins of modern science, on the basis of which I published Fuller (1997). I also designed the science and religion module for the M.Sc. in Science Communication at the Open University. In 2005 I was given the eye-opening opportunity to participate in *Kitzmiller v. Dover Area School District*, to which my home institution, Warwick University, responded with a measure of sympathetic interest which could serve as a model for how other such institutions should deal with their heat-seeking staff members. Finally, this book should be seen as twinned with another work that pursues in more detail the challenges to our understanding of humanity that both evolution and intelligent design pose (Fuller 2007: esp. Part III).

I am indebted to the following people for their input and generosity in providing opportunities, at various stages, for me to develop the ideas and arguments in this book: Thomas Basbøll, Michael Bérubé, Lyn Brierley-Jones, John Angus Campbell, Jim Collier, Pat Gillen, Kieran Healey, John Holbo, Kirk Junker, Bill Keith, Joan Leach, Mike Lynch, Chris Mooney, John Quiggin, Hans Radder, Zia Sardar, Mike Savage, Mark Smith, Nico Stehr, Rob Taylor, Jeff Thomas, Roger Trigg, and Maiko Watanabe. Special thanks also to Emma Longstaff at Polity Press for her persistence and patience, in equal measure. This book is dedicated to Dolores Byrnes, a woman whose nature defies description.

1

Historical Bases for the Problem

1. Science: Western Religion by Secular Means

Science and religion are not mutually exclusive categories. There is no evidence that belief in a deity, even a supernatural one, inhibits one's ability to study the natural world systematically. If anything, history provides evidence for the contrary thesis – that there is a synergy between the two. This is not to say that science and religion are identical. They are obviously institutionalized differently. Religion tends to permeate more of people's lives than science, so that people typically know more about the religion they believe than the science they trust. In any case, the conduciveness of a scientific theory to religious belief does not necessarily make it less scientific – nor does the conduciveness of a religious doctrine to scientific treatment necessarily make it less religious. Indeed, easy intercourse between science and religion ensures that science remains meaningful and religion avoids fantasy. The US pragmatist philosopher William James understood this point better than most.

When "Science" and "Religion" are capitalized as irreconcilable cultural forces, a specific sociological problem is referenced: *secular privilege for sacred knowledge.* In that respect, the supposed intellectual conflict between science and religion has really transpired as a political struggle between state and church. Outside this institutional setting, science and religion are difficult to contrast because the two concepts function in such markedly different ways. In particular, "science" is a positively marked and "religion" a negatively marked term. In other words, a look at the full range of things that count respectively as "science" and "religion" reveals that sciences are defined by what they are, whereas religions are defined by what they are not.

"Science" refers to the most authoritative form of knowledge over some domain of reality or the method used to obtain such knowledge. Thus disciplines struggle, often amongst themselves, to enjoy the title of science (Fuller 1997; Fuller and Collier 2004: ch. 4). In contrast, "religion" is a residual term that philologists and ethnologists started to use technically in the mid-19th century for complex systems of belief and, especially, ritual (the Latin root of "religion') that have not depended on the modern nation-state (Masuzawa 2005). To be sure, religions are often justified by appeal to a mythical history, in contrast with the explicit "social contract" by which Europeanized states have been founded and legitimated (Fuller 2006b: 134–6). But it does not follow that religions share any fixed, let alone "supernatural," attitudes toward reality, however abstractly one might wish to specify these. For example, world-religions like Hinduism, Buddhism, Confucianism, along with most tribal religions, however faulty their understanding of nature, do not require that its existence be underwritten by a supernatural, let alone personal, deity.

Matters are further complicated by the fact that science itself is a historically religious concept. Science may express globalization in its most distinctly Western inflection, but what marks this process as "Western" lies in its specific religious roots. In what follows I use the phrase "Western religious traditions," by which I mean Judaism, Christianity and Islam – the self-avowed successive moments in the spiritual quest of the biblical Abraham (Fuller 2006b: ch. 11). These so-called Religions of the Book trace the origin and character of reality to a single deity in whose "image and likeness" human beings were created, specifically in order to fathom the divine plan. Of course, virtually every culture has fostered a holistic understanding of reality, but typically reality is not portrayed as having been "created" by a deity with whom humans have a unique relationship that implies our privileged cognitive access and practical control of this reality – including perhaps a final convergence with the deity itself. This is the source of such popular images of physicists as "getting into the mind of God" and geneticists "playing God." Without the global ascendancy of this general world-view, it would be hard to explain science's motivation, ambition, pervasiveness, and persistence – especially when seen against the major dislocations and devastations that science has increasingly made possible.

Three general models have been proposed since the 19th century to explain the globalization of science, all involving some implicit attitude toward Western religious traditions:

1. This model is associated with the great philosophers of history of the early 19th century, Auguste Comte and G. W. F. Hegel, but also developed by the first professional historians of science in the early 20th century, Paul Tannery and George Sarton. It presumes that science is a unique development of Western consciousness, It amounts to a secular version of the Christian salvation story.
2. Championed by figures as disparate as Karl Marx and Thomas Kuhn, this model views the development of science as a potentially repeatable process proceeding through set stages, which at least to some extent is autonomous from ambient cultural influences. The phenomenon of "defensive modernization," whereby a historically non-Western country like Japan strongly institutionalizes science and technology without incorporating the West's political and ideological systems, fits here (Fuller 1997: ch. 6). The universality of science is the

knowledge which remains through iterations of cross-cultural translation, which typically involves the removal of science's original religious packaging.

3. This model concedes the most to the contributions of non-Western cultures, in particular concerning their access to knowledge (especially in the humanities and natural history) that would not be the normal part of Western experience, and that also may be specifically obscured by Western scientific categories. Western religious residues are often blamed for these blindspots. Yet, even here, non-Western forms of knowledge appear as merely complementing, not contradicting Western science, while lacking the West's global aspirations.

In contrast, non-Western cosmologies tend to portray humans as essentially embedded in nature, such that the deity is either coextensive with nature (i.e. pantheism) or transcends nature in a radically superhuman fashion that remains forever mysterious. These cultures have featured highly developed forms of learning and technique, which in the case of India and China often rivaled the West's until the late 18th century (Frank 1998). However, such knowledge has a domain- or problem-specific character or, as in the case of logic and mathematics, they are treated as pure discipline or technique (Collins 1998: chs 4–7). Little, if any, effort is devoted to reconciling disparate and contradictory bodies of knowledge in an overarching set of principles that might reasonably simulate a deity's blueprint. Thus, whereas India and China had their own versions of a "Renaissance man" like Leonardo da Vinci, who excelled at many different things, they never produced a single-minded "scientific revolutionary" like Newton. This was not through lack of talent but lack of *motive*.

This absence of science, even from very advanced non-Western cultures, should not be so surprising. Indeed, Westerners have recovered some of this sensibility in the postmodern condition, with its multiple configurations of "technoscience" that deny the very possibility of a unified systematic understanding of reality. The fact that the diffuse client-driven biomedical sciences have eclipsed the centrally state-funded discipline of physics as the paradigm of scientific activity is emblematic of this transition

(Fuller 2006b: ch. 12). It suggests that we need to step back and ask why *should* one aspire to a theory that unifies all things so as to enable the artificial reproduction and transformation of the natural order? Why not simply aspire to a theory that allows one to blend into nature with the least suffering to oneself and one's cohabitants?

The West's answer has been a belief that the natural order is the product of a single intelligence from which our own intelligence descends. This belief has produced, as the great 20th-century technological visionary Lewis Mumford (1934) put it, a "monotechnic" imaginary (cf. Noble 1997, for the dystopic version). Writing 150 years earlier, in the wake of the Newtonian world-system, Immanuel Kant went so far as to defend the belief as a necessary precondition for objective knowledge. And, while the Protestant Reformation inspired Newton to fathom the mind of God for himself without the mediation of priestly authority, it would be a mistake to identify this belief exclusively with Christianity. A crucial prior stage was the systematization and translation of ancient Greco-Roman learning into Arabic by Muslim scholars, starting in 9th-century Baghdad. For the first time in history, literary residues were organized not simply as cultural artifacts but a living legacy that spoke to universal human concerns. In this respect, the Muslim scholars conferred on the works of Plato, Aristotle, and so on, much of the same significance they had attributed to the sacred Jewish and Christian texts that were taken to have anticipated the Qur'an.

This attitude was conveyed to the European scholastics, once the Arabic translations themselves began to be translated into 12th-century Latin. It set in motion the project of synthesizing disparate bodies of knowledge into a higher-order unity. The project acquired an increasingly scientific cast, once God was portrayed as bound by the principles of his own creation, or "natural law." This idea, that God's perfection implies his having created the best of all possible worlds (and hence has no further need to intervene in it), was due to the greatest of the Muslim scholars, through whom the Christians learned of Aristotle in detail, the Spanish jurist Averroes. By the late 13th century, "Averroism" had become a Christian heresy that claimed precedence for the empirical study of nature over theological interpretation. Over the next three centuries, the University of Padua in Italy became a stronghold for Averroists, among whose students was one Galileo Galilei, the person who since the late 19th century has epitomized the tension between science and theology in the Scientific Revolution.

In non-Western cultures, it has been more natural to adopt a position that Westerners associate with the Aristotelian philosophical tradition – namely, that each kind of thing requires its own form of knowledge and that it is mistaken to impose a form of knowledge inappropriate to a thing's nature. The resulting cosmology implies a patchwork conception of reality, in which, say, mathematics is applied to some kinds of things but not others. That reality appears to consist of things understood in widely disparate ways is simply accepted, perhaps as a sign of our ultimate ignorance with which we should learn to cope, not try to transcend. Skepticism thus becomes a basis for Buddhist equanimity, not Cartesian anxiety. In this respect, Western science stands out for its treatment of nature not as a source of inherent value and cosmic order, but as an obstruction to the manifestation of our most god-like tendencies (cf. Noble 1997). These spiritual qualities were originally aligned with the "supernatural," but starting with Descartes and other 17th-century "mechanical philosophers" they were increasingly identified with the "artificial," as in today's prospect of "artificial intelligence" enhancing and perhaps even superseding human cognition.

To explain the distinctiveness of Western science is not yet to explain its global spread (Blaut 1993). Three factors have been historically prominent. The first concerns the relatively disorganized socio-economic conditions of Christian Europe in the Middle Ages vis-à-vis Islam and the more advanced Asian civilizations (Huff 1993). The second is related to the proselytizing character of especially Islam and Christianity. The third concerns a peculiarity of the organization of Christianity as a community of faith that easily generates heresies. This volatile feature was secularized and institutionalized as what Karl Popper regarded as science's unique method of "conjectures and refutations."

Crucial to the first factor is the observation that ancient empires operated according to a "tributarian" political economy, which means that the imperial power extracted revenue from the colonies, which it then either kept for itself or redistributed across the empire (Amin 1991). However, the local modes of production typically went undisturbed, which meant that there was little incentive for scientific and technological innovation. Such a system worked reasonably under conditions of economic and political stability. However, the European Middle Ages were marked by an increasing number of internal and external threats to the stability of the Roman Empire, even as it continued to expand. The power of the Roman bureaucracy yielded to the autonomous fiefdoms that characterized the feudal political economy, whose capacity to maintain peace and prosperity varied significantly across time and place. This patchy environment provided opportunities for local innovation and the free adoption, diffusion, and sometimes improvement of techniques from the more advanced Islamic civilization, as well as India and China. Moreover, whereas a technique like writing was under strict control of the elites of the great Eastern empires, and indeed functioned as a vehicle for maintaining social distance from the masses, such restrictions were impossible to enforce in Christendom, regardless of the desires of the Roman Catholic Church, which was plagued by wars of papal succession. Gutenberg's commercialization of the printing press in the 15th century was simply the last phase of this development (Febvre 1983).

In short, the "freedom of inquiry" valorized in Western culture that came to be a normative benchmark for science worldwide was historically an unintended consequence of ineffectual governance (Fuller 1997: ch. 5). Perhaps the decisive step in the delegation of power to local authorities was the introduction of the category of *universitas*, or corporation, in 12th-century Roman law (Huff 1993: ch. 4). The original corporations were churches, monasteries, guilds, and universities. They shared a concern with the maintenance and promotion of skilled practices whose value extends beyond the secular interests of a given group of practitioners. Such entities would be granted autonomy as long as they did not try to undermine the jurisdictions where they were

embedded. Thus, science was institutionalized as a legally pro-
tected "value-neutral" activity that could be conducted over
successive generations without interruption. This process would
be reinvented throughout the modern period with the granting
of royal and later national charters to self-governing scientific
societies. The overall effect was to constitute producers of knowledge as an independent class, whose mode of reproduction became
increasingly explicit (via formal curricula and examination) and
hence potentially open to all sectors of society, and its products
(via writing and technology) more easily diffused to disparate
social, economic, and political settings.

The second factor pertains to the proselytism inherent in the
Islamic and Christian roots of science, which in the modern
period became increasingly secularized and globalized via
Catholic missionaries (which brought Copernican astronomy to
China in the early 17th century at the same time as Galileo was
being persecuted for espousing it in Rome), Protestant evangelism, Enlightenment liberalism, European imperialism, and world
Communism. Science is not simply the knowledge developed by
and for the West. Rather, it is the knowledge by which *everyone*
may become fully human. It is against this backdrop that the
Western origins of science appear merely accidental (i.e. it could
have happened anywhere under the right circumstances because
everyone is equally competent). While it has been common to
stress the arrogant and even hegemonic qualities of this attitude,
the flipside is equally noteworthy: precisely because science is
supposed to be a common human legacy, the cultural origins of
a scientific knowledge claim are irrelevant to judgments of its
validity. Thus, science is distinguished by the increasingly multi-
cultural nature of its contributors. However, it remains unclear
whether this reflects the Westernization of non-Western cultures
or the de-Westernization of a truly global science.

For many scientific proselytizers, especially positivists and
Marxists in the 19th and 20th centuries, no clear line could be
drawn between the advancement of science and the scientific
reorganization of society. Moreover, as science spread to societies
lacking in the legal protections associated with the *universitas*,
scientists came to see their activities as potentially at odds with

those of their host cultures. Modeled partly on the captive "community of faith" that characterized the early history of the Jews and the Christians, and partly on the virtual "republic of letters" that kept autocrats in check during the Enlightenment, scientists across the world from roughly 1870 to 1970 strove to internationalize scientific communications, ostensibly to facilitate the unification of scientific knowledge, but equally to serve as a bulwark for protecting scientists from patriotic pressures that might compromise the integrity of their work (Schroeder-Gudehus 1990). Perhaps the most famous organizer of scientists as an international social movement was the X-ray crystallographer, John Desmond Bernal (1939), a British Marxist who viewed scientists as highly skilled members of the proletariat who, once organized, could provide the vanguard for social planning on a global scale. Bernal's efforts were not helped by his unflagging support for the Soviet Union during the Cold War, not to mention the increasingly divergent career patterns and ideological loyalties of scientists (Werskey 1988).

Finally, science has inherited and rationalized Christianity's propensity to heresy (Fuller 2003: ch. 10). In this respect, Galileo's challenge to the authority of the Roman Catholic Church was distinctive only in introducing alternative sources of dissent – especially as produced in artificial experimental environments and through the technological mediation of the telescope – to the time-honored practice of juxtaposing contradictory readings of canonical texts. To be sure, Galileo's signature brand of heresy was equally condemned by many key Protestant leaders who had themselves been branded as heretics. However, non-aligned Christians like Descartes, Hobbes, Boyle, and Newton saw in Galileo's challenge the potential for a kind of authority that could legitimately claim for itself the universality to which the Church of Rome aspired. Others at that time, notably Francis Bacon, the English Lord Chancellor (i.e. the King's lawyer), saw a special role for the emerging nation-states to adjudicate science-cum-religious claims by introducing a secular version of the Papal Inquisition, namely, the "crucial experiment" that could decide between competing hypotheses. This is the basis of science's "demarcation problem," and is discussed, now as a current legal issue, in chapter 4.

The word "heresy" comes from the Greek for "decision." Heresies are generated because potential converts to the Abrahamic religions are required to bear personal witness to God, yet the sincerity of that witness is usually not sufficient to secure acceptance into a community of faith. In this respect, an individual's decision to believe can be later collectively deemed to have been false, which then makes the believer a heretic, if he or she persists in the belief. Needless to say, this is not a recipe for tolerance, but it does take seriously the idea that ultimately, having been created in the image and likeness of the same God, we are all aspiring to the same truth, and hence it matters whether we believe the same things. Among the three great Abrahamic faiths, Christianity enjoys a unique position. On the one hand, it is distinguished from Judaism by encouraging proselytism, but on the other, unlike Islam, Christianity has historically lacked the political capacity to enforce doctrinal uniformity.

Unsurprisingly, as the historic seat of Christendom, the Eurasian region (including the Near East) has been the world's crucible for protracted conflict over *ideas* (as opposed to hereditary entitlements), as people are regularly encouraged, if not forced, to "justify" themselves, to recall a piece of Protestant jargon that has gradually migrated, courtesy of Kant, to the common parlance of modern epistemology. Galileo's admirers, especially Descartes and Bacon, thought that the methodical process of validating scientific knowledge claims could sublimate these antagonisms. The record here is mixed: science has undeniably escalated the global capacity for both innovation and destruction in the modern era. Is that anything but an unwitting extension of capitalism's "creative destruction" of markets? In any case, the West continues to stand out in both its generation and resistance of ideas that challenge received opinion.

2. Science vs Religion or State vs Church?

By the late 18th century, the high-water mark of the European Enlightenment, it was common to conceptualize theology as proto-science in the process of being superseded by science's full empirical and mathematical resources. A half-century later, Auguste Comte turned the tables and projected a "positivist polity" tantamount to a global religion founded on firm scientific foundations. While in retrospect this sequence looks like the basis for today's struggles between science and religion, it is better seen as successive stages in a much more specific process, the secularization of Christianity. In this context, "science" gives forward momentum to the decentralization of religious authority that had begun in the Protestant Reformation by stressing "reason" and "intelligence" as something that is possessed not simply by God, let alone theological experts, but by every human being, precisely because we are created in the image and likeness of God. Thus, common to this line of thought, which in the 19th century extended beyond Comtean positivism to Hegelian brands of socialism (e.g. Marxism), is the idea that individuals newly awakened in their rational capacities would freely agree to engage in a collective project to consummate their common humanity, which would be tantamount to realizing a "Heaven on Earth" (Becker 1932). In short, this version of the science–religion conflict was really a struggle against illegitimate authority in society at large: "science" stood for democratic rule, and "religion" autocratic rule. To be sure, the struggle had relatively little to do with the actual content of beliefs, even concerning God's existence.

A good way to appreciate the nature of the difference between science and religion is to imagine the narrative structure of two books, one called *The History of Science* and the other *The History of Religion*. The title of the latter book would immediately arouse suspicions that it is really a triumphalist account of The One True Faith, probably Protestant Christianity, as the last and least ritualistic (hence "purest") of the great world-religions. *The History of Science* would not raise such concerns because that book would be expected to explain the ascendancy of the sciences that are generally accepted as true across the world. To be sure, *The History of Science* would attend to the variegated sources of scientific insight – from scholars, priests, naturalists, astrologers, explorers, craftsmen, entrepreneurs, and so on – but the driving force of the narrative would show how these insights were mainstreamed into a common body of knowledge to which others from still more varied backgrounds then contributed. Here, ongoing philosophical reflections on the "scientific method" have played a key role in analyzing and synthesizing these disparate contributions to constitute "science" as a cross-cultural, trans-historical yet ultimately unified activity (Fuller 1997: ch. 3).

There is no comparable "religious method" as a plot device available for *The History of Religion*. Of course, the Abrahamic "Religions of the Book," as well as Buddhism, have become world-religions – but decidedly *not* by some synthesis with other religions into a more comprehensive body of knowledge. On the contrary, the modern movement perhaps most explicitly dedicated to this project, *theosophy*, looks to adherents of more conventional religions like a pastiche, cutting and pasting different features of different faiths, joined together by what theosophists regard as the common search of all religions for an apprehension of the Absolute. In practice, this means the cultivation of higher intellectual powers – including telepathy, clairvoyance, channeling – that its detractors deem parapsychological pseudo-science. However, it would be a mistake to say that theosophy has not generated academic interest. (One contemporary philosopher who takes seriously theosophy's self-understanding is the self-styled "critical realist" Roy Bhaskar (2000), who postulates that a mind that robustly exceeds its material origins is a necessary precondition for the conduct of science.) I mean here simply to observe that, whatever its merits, theosophy does not propose to discipline existing religions to enable them to pursue their supposedly common spiritual quest more efficiently – which is what one would expect of a "religious method" historically comparable to the "scientific method." Theosophy is not a "meta-religion" comparable to, say, positivism's "metascience": it is simply an alternative, "new age" religion.

Precisely because science and religion are such different *kinds* of concepts, there is no conceptual problem with saying that science might be conducted in a way that reflects or is influenced by religious attitudes. Here I mean not only, say, scientific creationism or theistic evolutionism, but also "scientism," which is not unfairly seen as a family of attempts to convert science into a secular religion (Fuller 2006a: ch. 5). Of course, it does not follow that all religiously inflected science is good science but the mere presence of a religious attitude does not constitute a proper test of scientific value. The first and fiercest public defender of Charles Darwin's theory of evolution by natural selection, Thomas Henry Huxley, dealt with this matter very fairly in his 1893 Romanes Lecture, "Evolution and Ethics." Huxley observed that while Darwinism's naturalistic world-view had been anticipated by, say, the Greek atomistic and Epicurean philosophers, as well as Hindus and Buddhists, none were inspired to develop their insights into a full-fledged science. Indeed, they were discouraged, mainly because that very world-view had persuaded them that humans are simply temporary arrangements of matter, no different in kind from other such arrangements, and hence should strive to avoid suffering for the brief time we exist in our current form.

Huxley's point, which I believe is historically correct, is that the monotheistic religions, which elevate the status of humanity as "the image and likeness of God," have been primarily responsible for science as a detailed, comprehensive view of the cosmos – an articulated vision of the unity of nature perhaps best exemplified by Newtonian mechanics. For Huxley, the challenge of his age was somehow to preserve that intellectually ambitious spirit – which had led Newton to believe he could know the mind of God – as Darwin's much less arrogant view of humans came to be accepted more widely in society (Fuller 2006b: 141–3). While Huxley was a fully signed-up Darwinian, he believed that Darwinism's metaphysically diminished sense of humanity could easily discourage the pursuit of science in the future as it had in the past. No discussion of the relationship between science and religion in contemporary society can be truly honest if it does not keep Huxley's concerns firmly in mind.

The very image of a "culture war" focused so starkly on "science vs religion" only emerged shortly after Charles Darwin had such a popular success with *Origin of Species* (1859). The battle lines were drawn between allies of Darwin, who held that nature could be explained entirely without appealing to supernatural forces, and his opponents, who held that God was a necessary part of any comprehensive explanation of nature. This particular culture war has always been fought much more in the classroom than the research site. Thus, Huxley campaigned hard for the place of natural science in general education, the inclusion of science laboratories on university campuses, and the granting of university status to technical institutes like his own Imperial College in London.

Even intellectuals who went to their graves opposed to Darwin, notably John Stuart Mill, fought alongside Huxley. The main issue for them was less the status of theology as a form of knowledge than the requirement to take Holy Orders to teach at Britain's flagship institutions, Oxford and Cambridge. On a sympathetic interpretation, it would seem, then, this particular culture war was the final phase of the *de facto* separation of state and church in the UK. (Anglicanism remains *de jure* the state religion, to which US Episcopalians continue to pledge allegiance.) It is difficult to appreciate the animus that was directed specifically against the privilege that the Church of England enjoyed in the public sphere (including multiple seats in Parliament), even though Anglicans counted among the most effective opponents of slavery and poverty. Such hostility prevented Mill, author of the Victorian era's leading book on the scientific method, *System of Logic*, from ever meeting – let alone acknowledging agreement with – William Whewell, the most influential figure in Britain's scientific establishment, and indeed the person who coined "scientist" as the name of a profession (Snyder 2006). However, Whewell, Master of Trinity College, Cambridge, was also an ordained priest. In today's intellectual climate, it is ironic that not even the horror that both Mill and Whewell expressed at Darwin's renunciation of cosmic design was sufficient to overcome the difference between taking and not taking Holy Orders.

In any case, the last quarter of the 19th century witnessed the emergence of histories of science that projected this dispute back to the perennial problems scientists have faced from religious authorities. The titles of these books reflect the strong feeling that inspired their composition: John Draper's *History of the Conflict of Religion and Science* (1875) and Andrew Dickson White's *A History of the Warfare of Science with Theology in Christendom* (1895) were the leading representatives of this genre in Britain and America, respectively.

To be sure, a list of scientists who have been persecuted in the history of the West would include many religious heretics and nonconformists, but no one who explicitly denied the existence of God. In fact, "atheism" is a rather elusive concept in the history of Western thought – more something of which one has been accused than a mantle proudly assumed. A big problem has been simply a lack of conceptual and lexical resources for expressing a sense of existing without God (Febvre 1983). Indeed, until the 19th century, the situation was not unlike what social constructivists experience today (cf. Robertson 1929; Koertge 1998). On the one hand, they are accused of denying the existence of "reality" or "truth." On the other hand, of course, they cannot possibly deny such things because that would undermine the validity of their own knowledge claims! Consequently, social constructivists must be lying, bluffing, or committing some other intellectual felony that is already recognized by sincere truth-seekers and reality-huggers – and are prosecuted accordingly, with denial of tenure nowadays replacing burning at the stake.

Scientists from earlier eras like Roger Bacon (1220–92), Michael Servetus (1511–53), Giordano Bruno (1548–1600), and, last but not least, Galileo (1564–1642) ran afoul of Christian authorities by trying to reconcile their findings with their faith in innovative ways. By our lights, their problem was not so much one of "science vs religion" as of the freedom to inquire and teach within a common Christian culture. Their latter-day descendants are to be found among academics who would dare push the boundaries of science in our avowedly secular humanist culture: e.g. those who would defend a racial, or otherwise genetic, basis to intelligence, or an extension of moral and even legal rights to appropriately sentient animals (cf. Herrnstein and Murray 1994; Singer 1975). Just as we would not hesitate to identify today's heretics – albeit including such strange ideological bedfellows as the libertarian political economist Charles Murray and the philosopher of animal liberation Peter Singer – as secularists (though perhaps not humanists), likewise the earlier heretics are all Christians (though perhaps not Trinitarians).

In the case of the earlier set of figures, true to the etymology of *heresy* ("decision"), they asserted discretion over matters – such as the balance between textual and empirical authority – where the balance was presumed to be set (cf. Evans 2003). For example, among Galileo's sacrileges was his suggestion that had the Church Fathers enjoyed the benefit of peering through his telescope, they would have reconsidered the accuracy of the Bible's views on physical astronomy (Brooke 1991: 77–80). The Papal Inquisition would have been satisfied had Galileo simply split the difference between science and religion, granting science unparalleled utility in mapping the heavens and taming the Earth, while reserving for religion access to the ultimate nature of reality. Galileo, of course, notoriously refused to yield to this "double-truth" doctrine.

Heretics, both then and now, tend to hold views of reality that have been called "unitarian," "monistic," or "reductionist" – depending on whether the frame of reference is theological, philosophical, or scientific, respectively. They level knowledge-based hierarchies, effectively eliminating the expert basis for the exercise of power. The result is to force people who had previously kept a studied distance from each other's knowledge claims into direct competition, which has often resulted in conflict. Consider the heretical scientists mentioned in the previous paragraph who preceded Galileo: Bacon held that humans could take hold of the principles by which God created the world to arrive at creations of their own; Servetus proposed pulmonary circulation in the course of denying that the Holy Spirit exists independently of material reality; Bruno postulated multiple inhabited worlds by denying any principled distinction between a scientific understanding of the Earth and the heavens.

While in all these cases the heretics' flattening exercises aimed to give scientific inquiry the upper hand over theological orthodoxy, it is worth noting that the orthodoxy in question was the Roman Catholic Church (though Servetus was ultimately executed in Calvin's Geneva). The situation is exactly reversed in the more fundamentalist reaches of Islam and Protestant Christianity (Armstrong 2000). Indeed, the very idea of double, or even multiple, truths was originally advanced not by religious authorities but by those wishing to prevent potentially radical scientific

claims from either undergoing or undermining theological authority. This "Gnostic" motivation is traditionally associated with
Plato's influence on Judaism, Christianity, and Islam. In the 12th
century, Averroes, the Muslim philosopher largely responsible for
introducing medieval Christendom to the works of Aristotle, held
that free inquiry should be the province of an elite cult with the
intellectual wherewithal to remain faithful even after realizing the
Qur'an's literal incoherence. For their part, the masses should be
simply taught the literal but mysterious truth of the holy book.

Religious authorities at the time were scandalized not merely
by the Averroist view that God is bound by the laws he creates
(a doctrine designed to secure the autonomy of natural science
from theology but which, at the time, seemed to deny divine
omnipotence), but more importantly by the denial of the universalist idea common to the Abrahamic faiths, namely, that we
are all moral and cognitive equals in the eyes of God. But make
no mistake: the kind of universalism Averroes denied would
subordinate science to religion, not the other way round. Thus,
the doctrine of "double truth" with which his name has come
to be associated should be seen not as aiming to stratify an otherwise democratic regime but to liberalize an otherwise authoritarian regime by securing a specialist niche (Huff 1993: chs 2–3;
cf. Fuller 1997: ch. 6). Of course, what differs in doctrine may
not be so clearly distinguishable in practice, as currently illustrated by the West's conflicted attitudes towards the Islamic
Republic of Iran. On the one hand, Iranian scientific and technological achievements and aspirations cannot be denied; on the
other, neither can its subordination of the technoscientific
impulse to a religiously inspired democratic sensibility.

3. Converting the Divine Spark into the Scientific Pulse

"Creationism" is a term fraught with multiple meanings, some of
which are clearly irrationalist and anti-scientific, and others which
have been historically instrumental (and perhaps even conceptually

necessary) for the emergence and maintenance of rationality and science. Supporters of Darwinism tend to conflate the two so as to obscure creationism's rationalist side, a point to which we shall return in chapter 4, when considering the rhetoric of "supernaturalism." Creation narratives are, of course, common to most cultures. Usually the creative deity – even when portrayed in superficially personal terms – does not operate in a way that affords detailed and systematic understanding by those created, especially humans. Recognition of the qualitative difference stipulated between the creative power of this supernatural agent and that of human agents thus serves to instil an acquiescent and fatalistic attitude toward what one may expect or want of life. When so-called traditional or pre-modern societies are said to be "static," this is typically what is meant. The function of culture in such societies is to provide a supportive environment for cycling through the indefinitely repeatable stages of the life process.

However, creationism is subject to an epistemological step-change once the creative deity is portrayed as an intelligent designer. The deity becomes an entity whose creative powers differ only in degree not in kind from human creators: God is essentially a very big and very smart (and very good) engineer. Reality is clearly conceptualized as a *universe*, that is, the product of one hand that bears the distinctive intentions of its producer. This universe is typically conceptualized as an artifact, if not a machine – that is, the sort of entity that one would expect a human, but not an animal, to have produced. (Keep in mind that these claims were originally honed long before the advent of evolutionary psychology!) In that case, reciprocal inquiries between the "outer" world of nature and the "inner" world of mind should be mutually reinforcing. Both archeologists and explorers of extraterrestrial intelligence routinely engage in this form of reasoning (Ratzsch 2001).

If nothing else, the idea of intelligent design implies that reality is fruitfully regarded as mind-like in its construction, what philosophers dub the "intelligibility of nature" (Dear 2006). Psychology and physics thus become two sides of the same coin, and time-honored regulative principles of scientific inquiry, such as the appeal to "parsimony" or "economy" (i.e. explain the most by the least), make sense as design features of the universe. Appreciation of this duality informed how experimental psychology arose from "psychophysics" in mid-19th-century Germany and continues to inspire the "anthropic principle" in cosmology, whose supporters include the Nobel-Prize winning physicist and avowed atheist, Steven Weinberg (2001). In our own day, when the biochemist Michael Behe (1996) claims the "irreducible complexity" of the cell or a part of an organism as evidence for intelligent design in nature, the main appeal of his argument is to the made-for-purpose character of such things. This implies that God's creative capacities are sufficiently like our own that we can understand, say, the cell as if we had created it ourselves.

This leaves open the question of whether we can *re-engineer* nature as well: is it possible to strategically alter, if not outright improve, the functioning of an organism, or one of its parts, by replacing or even eliminating something with which it was born? This question was already posed in the 1880s by German biologists who had developed experimental protocols for interfering with the development of embryos of sea creatures like urchins and frogs (Cassirer 1950: ch. 11). The big lesson to take from this line of research – including its intensification at the genetic and molecular levels in the 20th century – is that claims about the irreducible complexity of a specific biological entity may stand or fall without necessarily affecting the general thesis of intelligent design. Take Behe's own favorite example of an irreducibly complex entity: the mousetrap. A broken mousetrap may either still function as a passable mousetrap or serve some entirely different function. Why cannot these be treated as contained within the design of the mousetrap (as what philosophers call "dispositional" properties) that are only realized under certain conditions, perhaps requiring the collaboration of some other (human) agent?

By way of contrast, a belief in the *radical* difference between divine and human creation should incline one to reason that God moves in such mysterious ways that it is not surprising that the chance-based processes of Darwinian evolution by natural selection would explain the constitution of cells and organisms. Perhaps this is why some scientists who are Christians, especially Roman

Catholic followers of Thomas Aquinas, have been keen to embrace Darwin – precisely to enable an ultimately unfathomable God to travel through the conjoined improbabilities required of natural selection. A notable case in point is Kenneth Miller (1999), co-author of the best-selling US high school biology textbook and star witness for the plaintiffs in *Kitzmiller v. Dover Area School District*, the latest legal wrangle between evolution and intelligent design – more about which in chapter 4.

For the intelligent design theorist, then, science is not unfairly seen as systematically second-guessing the deity, in which our understanding of "how" things are is a prelude to making sense of "why" they are as they are. In short, things happen or exist for reasons we may have yet to determine, as we fathom the mind of God. In the end, however, knowledge of mind and nature should be unified, reflecting the structure of the universe. Immanuel Kant enshrined this sensibility as the hallmark of the modernist world-view in the *Critique of Pure Reason* (1781), a work that aimed to justify *not* belief in God's existence per se, but belief in God's existence *as a necessary precondition for the rationality of the scientific enterprise*. Kant had no illusions about his ability to prove God's existence, but equally he realized that if we treated God as an illusion, we would lose the main reason for continuing to do science.

In this context, it is important to distinguish science from technology. For example, the Chinese empire, the world's economic superpower until the early 19th century, managed to make many technological advances without ever having staged a "Scientific Revolution" (Cohen 1994: ch. 6). From the standpoint of visiting Europeans in the 17th and 18th centuries, the Chinese did not appear to make the most of their innovations. Unlike the Europeans imbued with a divine mandate to create a "Heaven on Earth," the Chinese refused to see nature as mere raw material to be molded to human ends (cf. Noble 1997). At first, it was common to diagnose this situation as emblematic of what Hegel and Marx called "Oriental Despotism," whereby the Chinese reverence for nature masked an authoritarian ideology designed to inhibit the self-realization of the mass of humanity (Dorn 1991: chs 1–2). However, as Darwinism has spread across the globe over the past century (including an enthusiastic welcome in China), the Chinese avoidance of the Scientific Revolution has come to be interpreted more generously, as marking an ecological ethic from which the West may now, belatedly, learn.

While scientific creationism is most clearly rooted in the Abrahamic faiths, according to which The One True God appears to address only humans, it can also be found in ancient Greek thought that would have to count as "pagan." Indeed, the ancient Greek word for the natural order, *cosmos*, was originally taken from architecture. The analogical links between designing a temple and the universe are evident in the pre-Socratic philosophers, especially Anaximander (610–546 BC), for whom science and religion were mutually reinforcing activities (Hahn 2001). The original analogy still resonates today in philosophical talk of "foundations" for knowledge, even when philosophers do not mean to suggest the existence of a "founder." But does it make sense, even metaphorically, to refer to "foundations without a founder" or "design without a designer" or, for that matter, "selection without a selector"? This question comes very much to the fore with the ascendancy of the Darwinian doctrine that chance mutations are the driving force of evolution (Menuge 2004).

However, as intimated in our earlier discussion of Averroes, the challenge was originally posed the other way round: the acceptance of free scientific inquiry might render strictly theological discourses about the creative deity nugatory, unless science was kept from public view. Consider this personification of the problem: once an omnipotent, omniscient, and omnibenevolent deity had created the best possible world, what could the deity do that could not be grasped by the systematic study of nature? Indeed, what could the deity do *at all*? Moreover, why read the Bible to gain obliquely through poetic imagery what could be perceived directly through the senses? All told, the Averroists appeared to portray the divine disposition as *deus absconditus*, God fleeing the scene of creation after completing (at least his end of) the job. This view became prominent during the Scientific Revolution in the 17th century and characterized the self-styled "deistic" thinkers of the 18th-century Enlightenment.

The basic strategy to keep the theology relevant to the conduct of science in the face of the Averroist legacy has been to reassert the *voluntary* nature of divine creation. This move had already been present in the early days of Islam in the *kalam*, a type of rational theology that stressed a vision of God similar to that of *logos* in the versions of Christianity that were strongly influenced by Greek Stoic philosophy (Collins 1998: 451–5). The salient point about the *kalam* is that God is a self-caused creator who decides, moment to moment, to extend the law-like structure of creation through time, the implication being that at any given moment God could choose, by miraculous intervention, to suspend natural law and then later to reinstate it. That God may find it better not to do so is itself merely a fact of divine will, not a compulsion by natural law.

The profound impact of the *kalam* on the conduct of science in the monotheistic religious traditions cannot be overestimated (Wolfson 1976). Today it underwrites the idea that creation itself was a divine choice naturalistically expressed as the "Big Bang" hypothesis in physics, or perhaps the so-called Cambrian explosion, whereby the blueprints for most of the Earth's life forms came into existence roughly together 500,000 years ago. But, more subtly, the *kalam* extends the psychological range of the biblical claim that humans are created in the image and likeness of God. In particular, our proven cognitive capacity to imagine alternative hypotheses compatible with the same data is seen as grounded in God's ever-present creative capacity to choose from among many internally coherent possible worlds. This putative relationship between the divine and the human intellects has led to conclusions that have been both empowering and incapacitating for science.

A good example of this ambivalence that in its day proved incapacitating appears in the 14th-century Christian philosophers, William of Ockham (of "Ockham's Razor" fame) and Nicholas of Oresme, who together anticipated most of the ideas associated with the Scientific Revolution that would get into full swing only 300 years later. They were sensitive to the misleading character of language in trying to get at the nature of things. They seriously questioned the then-dominant Aristotelian idea that nature was essentially as it normally appeared. They countenanced the extension of quantification into realms that Aristotle denied in principle. They demonstrated that human understanding is always relative to its frame of reference, such that, appearances to the contrary, the Earth itself may be moving through space along with the other planets. And so on. Yet, in the end, Ockham and Oresme did not trigger the Scientific Revolution because they remained skeptical about the prospect of these matters ever being resolved – including by experimental means. For all their surface radicalism, Ockham and Oresme still drew a clear line between the human and the divine intellect. Thus, in this version, the *kalam* served to inhibit the development of science in its modern technologically enhanced sense.

So how, then, to navigate scientific inquiry through the Scylla of a God who creates so well at the outset that his own presence is made redundant and the Charybdis of a God whose arbitrary interventions potentially subvert sustained efforts at comprehension? The one undermines the need for theology, the other the possibility of science. As we have seen, both Muslim and Christian thinkers struggled with this dual prospect afforded by the Hebrew Scriptures. The answer that came to the fore in the early modern era was *theodicy*, a hybrid of theology and science (what we would now call a "philosophy of history") that portrayed the events as unfolding toward a perfectly rational whole that would be fully comprehended at the end of time (Milbank 1990). From the standpoint of theodicy, rational inquiry in its purest form is tantamount to the search for divine justice: why do things happen as they do? The biblical source for this question was the need to explain the centuries of Jewish captivity depicted in the prophetic books of the Old Testament: what might superficially appear as God's inability to protect his chosen people, the prophets interpreted as failed tests of faith from which the Jews must learn in order to become one with God (Funkenstein 1986: ch. 4).

The most salient phenomena for theodicy were events lacking any *prima facie* rationality: monstrous births, premature deaths, natural catastrophes, inhuman cruelty, and so on. Such events serve as reminders of the gap in human comprehension of the divine plan – but equally as divine intimations of the universe's intelligent design. Any apparent evil will bring forth, in the fullness of time, a good that could not have been produced more efficiently. It was just this view that Voltaire satirized in his celebrated Enlightenment novel, *Candide*, in which Dr Pangloss easily interprets any disaster as a blessing in disguise. That theodicy's wittiest scourge, Voltaire, was also Newton's staunchest French defender is no accident, since theodicy was Newtonianism's main competing strategy for harmonizing science and theology. In the next chapter, we shall consider the Newtonian legacy to intelligent design, but let us briefly consider theodicy's.

Modern theodicy dates from Calvin's revival of St Augustine's "providentialist" view of history, which justified perseverance in faith during times of hardship, the full benefits of which would not be redeemed in one's own lifetime but only at the end of time (Löwith 1949). Theodicy enjoyed adherents across the early modern religious spectrum, ranging from Leibniz, who promoted a reunited Christendom in the wake of the Reformation, to Bossuet, whose loyalty to Louis XIV was underwritten by the divine right of kings (ch. 7). Over the next two centuries, theodicy was subject to *activist* and *passivist* secularizations: the former increasingly militant and apocalyptic; the latter quiescent and fatalistic. The activist tradition includes Vico, Condorcet, Comte, and Marx. The passivist tradition includes Hume, Malthus, Spencer, and Darwin. Some figures – notably Adam Smith and his early German reader G. W. F. Hegel – straddle the left–right political divide but remain very much wedded to theodicy's "cunning of reason" mentality (cf. Fuller 2006a: ch. 5). We shall revisit these two strands of secular theodicy in the next chapter, as the activists harbor silicon dreams of humans evaporating into the ether and the passivists see our future inextricably bound with all the other carbon-based creatures on planet Earth.

Theodicy's appeal lay in its assumption that every event is "meaningful," in the sense of necessary to enable humans to grasp the divine plan in its totality. History is not wasted time, secular delay in the face of sacred inevitability, but rather the process of humanity's collective maturation. Hegel famously grasped this point at the dawn of the 19th century when he defined history as philosophy – in the previous century, he would have said theology – teaching by example. For many, theodicy seemed to strike the right balance between human arrogance and humility – in contrast to Newtonianism, whose reductionist metaphysics harked back to the alchemical dream of mastering the formula of creation for purposes of improvement, implying that God did not do the job right the first time or cannot finish the job without the aid of humans. Needless to say, it is easy to see how Newton's "arrogant" theology, which he only partially succeeded in suppressing in his lifetime, would end up converting science from a purely intellectual activity to the premier technology of globalization (Noble 1997). For its part, theodicy has left its mark on modern theories of knowledge, albeit in a much diminished state: It explains why epistemologists feel compelled to demand "justification" for our beliefs as evidence that we hold them for good reason, regardless of their ultimate truth. Just as God insists that each religious believer find his or her own way to the truth, so too the epistemologist requires this of the secular believer.

4. The Two Faces of Design in Creation and Evolution

The real intellectual differences between evolutionary and creationist accounts of life are much harder to pin down than the rhetorical heat generated by their conflict would suggest. Darwin himself is not unreasonably regarded as a recovering creationist. Certainly this would explain his resort to a phrase like "natural selection," which conjures up the idea of "design without a designer." To the discerning ear this may sound like someone whose mastery of alcoholism amounts to the compulsive drinking of non-alcoholic beer. While this is not the place to consider why Darwin retained the language but not the ontology of design, nevertheless it leads one to suspect that whatever genuine differences may exist between evolutionists and creationists, they may be overshadowed by mirror-image differences that exist within both camps.

The sorts of differences I have in mind may be epitomized by two slogans about the nature of *design* itself:

1. *Form follows function*: Physical structures survive because they serve a function in some larger system. Different structures may better serve those functions at other times and places.
2. *Function follows form*: The functional specification of a system is constrained by the available physical structures. Absent the appropriate structures, and certain functions simply cannot be performed.

These two slogans are nowadays most frequently heard in schools of "design," in the sense of architecture and the built environment. Here (1) is heard as the modernist motto associated with the Bauhaus school that flourished in the German-speaking world in the interwar period, whose influence extended to the logical positivists (Galison 1990). Such "functionalists" typically saw themselves as advocating a design strategy focused on a normative conception of "life" that could be realized in better or worse ways in natural and artificial worlds. In contrast, depending on ideological orientation, (2) may stand for either a pre- or a post-modernist position, both of which shift the design emphasis from the motivating end to the available means. In the case of pre-modernists, these means are often epitomized as "tradition" or "legacy," whereas postmodernists tend to stress the material residues or unintended consequences of any intentional act that then provide the conditions for any further action.

Adherents to these two contrasting slogans can be found amongst both creationists and evolutionists. Take creationists first. Some like Newton adopt the standpoint of the creator, thinking in terms of a blueprint that God laid down at the outset that provides at least the outer constraints (i.e. the laws of physics) within which all possible life must occur. This is a clear case of function following form. However, others like William Paley, who so influenced Darwin, shift from God's original decision about blueprints to the function served by particular organisms being as they are. The former stresses the "underdetermination" of creation (i.e. God might have created other than he did); the latter the "overdetermination" of creation (i.e. whatever God created was created for a reason). While the two positions provide complementary perspectives on the same phenomena, neverthe-

less historically they have pulled in opposing ideological directions. Newton's view inspired considerable technological activism,
whereby humans harness the power of physics for their own ends,
whereas Paley's promoted a more quiescent sense of ecological
stewardship that aims to preserve nature's equilibrium as God
presumably intended it.

A useful way to think about the general difference between
creationist and evolutionist perspectives is in terms of where each
locates God. The creationist deity, as we have seen, clearly transcends nature: God infuses an inherently unintelligent ("disor-
dered") nature with intelligent design ("order"). This was the
original sense of the word "information," i.e. God "informs"
matter by giving it a rational structure. It also fits the general
spirit of the Genesis creation story. In contrast, the evolutionist
deity is immanent in nature, indeed, perhaps to such an extent
that the evolutionist finds it unnecessary to refer to God as a
separate entity at all. As we shall see in chapter 4, this was the
original sense of "naturalism," a term first used pejoratively to
refer to the ideas of the late 17th-century philosopher Spinoza.
This resulting image is suggested by the persistence of paradoxical
phrases like "natural selection" and "design without a designer":
For the evolutionist, nature appears to acquire God-like qualities,
typically when one shifts from a micro- to a macro-perspective.
The implications of this image can be either pessimistic or
optimistic, but in a way that reverses the ideological polarity we
saw operating with the creationists.

On the one hand, theologians enamored of Newton liked to
demonstrate God's rationality and freedom by imagining that the
deity mentally discarded other possible but less perfect plans for
the universe before selecting the plan he actually implemented,
whereas Darwin's God discards his blueprints openly over time
in the form of extinct species, thereby demonstrating blind and
wasted effort rather than forethought. Indeed, one point on
which Newton and his great contemporary antagonist, Leibniz,
agreed was that God largely created what Paul Davies (2006) has
recently called a "Goldilocks universe," whereby the universe's
physical constants are set "just right" for humans to be able to
fathom its structure (aka the anthropic principle), while the degree

of death and destruction exhibited in nature led Darwin to doubt the existence of God altogether – and perhaps even whether humans would ever understand the cellular and sub-cellular levels of life.

On the other hand, God's absorption into nature effectively gives nature a mind of its own, one striving to realize an end that ultimately requires a major metamorphosis of species. This more upbeat vision of evolution depends on a much more permeable boundary between organism and environment than is presumed in Darwin's own downbeat version. In other words, the environment relates to organisms in a more nuanced fashion than the simple life-or-death (or, more precisely, reproduce-or-not) judgments of natural selection. In particular, the environment retains much of the missing deity's creative capacity to shape organisms over time to improve their fitness. This was the perspective of Jean-Baptiste Lamarck, who put forward the first modern scientific theory of evolution. We shall see more about him below and throughout this book. Lamarck is a figure of recurrent fascination because his was an unabashedly anthropocentric, progressive view of evolution that treated all earlier life forms as anticipations of our best possible selves, a fate that is bound to be realized as we take greater control over the course of evolution and, in Darwinian terms, provide sight to the blind watchmaker (cf. Dawkins 1986).

Based on the preceding discussion, two strands in the history of science may be identified that correspond to the two design slogans. Their basic features are epitomized in Table 1.1. Perhaps the most basic way to distinguish these two historical strands is in terms of the sort of entity they track over time. Those in the "form follows function" tradition track *functions*. They begin by looking at a living organism, imagining their parts to contribute to the overall functioning of the whole, and then ask how would the same or comparable functions have been performed by the parts of other organisms in other times and places. One can then inquire into the effectiveness of their function and perhaps even speak of progress over time. This mode of analysis is also common in the history of technology. The very idea of a "history of transport" presupposes a functionalist standpoint because one

Table 1.1 The two design traditions in the history of science

Design slogan	"Form follows function"	"Function follows form"
Principle of design	Analogy of functions	Homology of forms
Creationist spin on slogan (What do humans do?)	Contemplation of plan	Completion of plan
Evolutionist spin on slogan (What do humans observe?)	Increased effectiveness of life (degree of adaptation)	Increased waste of life (degree of non-adaptation)
Creationist science	Natural theology	Natural philosophy
Evolutionist science	Physiology	Anatomy
Conception of historical change	Uniform gradualism	Punctuated equilibrium
Creationist icon	William Paley	Isaac Newton
Creo-Evo icon	Jean-Baptiste Lamarck	Georges Cuvier
Evolutionist icon	Richard Dawkins	Stephen Jay Gould

imagines at the outset that people have always wanted to move themselves and other things back and forth, and that various technologies may be compared with regard to how much can be moved how far by what means to what effect. These technologies – including horses, wagons, boats, trains, cars, planes – vary considerably in material composition and structural principles, but have performed sufficiently analogous functions to be sensibly included in a common history of transport. A plot device in such histories, which one also finds in branches of psychology and economics, is the never-ending quest for more "efficient" or "productive" means of achieving the same or enhanced ends.

In contrast, the "function follows form" tradition tracks *forms*, whose functions may change over time. Stephen Jay Gould marked this key difference from functionalism by coining the term "exaptation" to mean a physical structure that begins by serving one function but that, over time and in consort with other changes in the organism and the environment, comes to serve an entirely different function (the intended contrast is with the functionalist's fixation on "adaptation" as the mark of evolutionary survival). To be sure, this view is to be understood as

assuming that if an organism and/or the environment changes
sufficiently, then the physical structure in question may simply
not survive at all. In that respect, the beginning of the evolutionary narrative may provide little indication of how it is to end.
This general line of reasoning is also familiar from anthropology
and sociology, where physical parameters, especially population
growth, density, and longevity, can increase or diminish the
chance that certain social functions will be performed in, say, a
dedicated fashion by a specialized institution. In this context, one
often speaks about the "unintended consequences" of social
action, whereby increased activity of a certain sort proves over
time to be self-defeating, as they provide conditions for new
practices that crowd out the original ones.

In chapter 3, I shall examine in more detail the difference
represented by Dawkins and Gould in the final row of Table 1.1.
However, in what follows, I will focus on the "Creo-Evo"
moment, which occurred around 1800 and marked the reversal
of the creationist and evolutionist interpretations of the two
design slogans. Lamarck and Cuvier were rival Parisian museum
directors (of, respectively, invertebrate and vertebrate animals) in
the first period of systematic specimen collection of extant and
extinct organisms. These activities were motivated by state-driven
economic expansion, both abroad and at home. Foreign expansion resulted in the drive to systematize all of life under a common
set of principles that persists to this day, while domestic expansion
explained the mass excavations (for mines, roads, factories, and
houses) that revealed a previously unknown fossil record (Pyenson
and Sheets-Pyenson 1999: chs 5–6).

Two general phenomena impressed themselves on those who
lived through this pivotal period: the diversity of life and the
pervasiveness of death (Toulmin and Goodfield 1965: ch. 8).
These would provide the empirical cornerstones of Darwin's own
theory of evolution, a half-century later. But at the time they
simply cast doubt on any straightforward account of God's creative
activity in nature. The stratified nature of the Earth's geology gave
the appearance of a stumbling God who lays down successive
drafts of nature's plan, while the creatures brought into existence
in the plan's latest (final?) draft display a bewildering variety that

confound attempts to uncover Newton-like laws of biological design. In particular, despite their differences, Lamarck and Cuvier worried that the evidence pointed away from the idea of "common descent," i.e. that all life had descended from an original creative moment. However, Darwin would revive a secular version of "monogenesis" in terms of a common biological ancestor – in the famous image of Darwin's leading German defender, Ernst Haeckel, who refashioned Francis Bacon's tree of knowledge (which justified the disciplines as functionally differentiated outgrowths of the human mind) into the "tree of life" still found in secondary school biology texts, which presents all of life as having evolved from a seed-like single-celled organism.

Despite a common Catholic background – Lamarck a Jesuit-trained dissenter and Cuvier a more traditional practitioner – they were inclined to let God speak entirely through nature, though without the deity quite becoming identical with nature. They sidestepped the question of life's origins. But Lamarck went one step further, notoriously arguing that, regardless of origins, nature's intelligence is exhibited by the gradual winnowing away and convergence of life forms in an anthropomorphic direction. Thus, for Lamarck, the eligibility of "biology" (his coinage) for the status of science lay in life possessing a unified nature – as creative intelligence – that appears in a variety of disposable physical forms (aka species). In the pages that follow, we shall see that this idea has powerfully gripped both the sociological and biological imaginations, including sociology's founder, Auguste Comte, St George Mivart, Darwin's fiercest critic among the biologists of his day, and Julian Huxley, the modern theorist of evolutionary humanism. Indeed, the convergentist sensibility is currently enjoying a revival of sorts, albeit without explicit reference to Lamarck, as geneticists come to realize that structurally disparate creatures appear to perform similar functions by virtue of possessing common genetic arrangements (Carroll 2005: 71–2; cf. Morris 2003).

The Creo-Evo moment originally occurred against the strongly anti-clerical climate of post-revolutionary France. It is worth observing that after this initial crisis of confidence, creationism periodically received a renewed empirical lease of life over the next two centuries. Two major moments stand out. First, in the 1830s, William Buckland, Whewell's Oxford counterpart, argued that the emerging paleontological evidence pointed to one creative moment in the distant past when the prototypes of most current life forms came into being (Knight 2004: ch. 5). To be sure, Buckland stuck to biblical chronology of creation and flood when he identified the Cambrian explosion of a half-million years ago, followed by the Permian extinction a quarter-million years ago, when 96 percent of all species died (Gould 1989). Nevertheless, the empirical side of his observation remains intact. Even the idea of a relatively young Earth persisted through the 19th century – not as young as the biblical creationists would like (a few thousand years old), but young enough (a few million rather than a few billion years old) to make it unlikely that evolution occurred by the chance-based processes that Darwin's theory of natural selection required. This all changed, however, in the early 20th century, with the discovery of the decay rates of atoms, which led Ernest Rutherford to suggest that the long half-lives of recently discovered radioactive elements enabled their traces on rocks to be used for measuring geological time.

Creationism's fortunes were once again revived with the incorporation of genetics and molecular biology into what is now called the neo-Darwinian synthesis in the 20th century. I shall have more to write about the creationist roots of genetics in the next chapter, but for now suffice it to say that proponents of the latest version of scientific creationism, intelligent design theory, make much of the fact that Darwin himself seemed to believe that the microstructure of organisms, including cells and genes, would remain forever mysterious, which contributed to the overall "blindness" of evolution, over which humans could never expect to gain definitive control (Behe 1996). Although today's evolutionists have (so far) scored rhetorically in portraying intelligent design theory as a "science-stopper" because of its alleged recourse to God as an explanatory principle, ironically Darwin's own principle of natural selection was seen by critics like Whewell and Mill, the leading philosophers of science of their day, as just such a science-stopper for undermining the intelligibility of nature.

Darwin fundamentally resisted the idea that nature possesses a machine-like structure that might imply, as it did for Newton, a divine mechanic. This came out in his inability, if not refusal, to fathom the mathematical formulae that purported to provide evidence for statistical laws of heredity, sent to him by Gregor Mendel shortly after the publication of *Origin of Species*. As it turned out, successive developments in largely laboratory-based work in the 20th century reduced the semantic distance between "natural selection" and "artificial selection." What Darwin had proposed as a metaphor between how nature and a farmer would selectively breed plants and animals turned out to be much closer to an identity. The cracking of the genetic code in the second half of the century revealed that behind the radical differences in structure among plants and animals was a 90+ percent overlap of genetic material, plus a few amino acids arranged in distinctive ways.

Moreover, our increasing success at genetic engineering, and biotechnology more generally, points to our growing competence in *replacing* nature in the "selection" process. Indeed, we may soon need to take the throwaway phrase "playing God" literally. A quarter-century ago, the founder of modern bioethics, Jonathan Glover (1984), had already argued against any presumptive preference for promoting traits just because previous generations of humans had been born with them: "human nature" is simply shorthand for past burdens, which only partially constrain our creative agency. Such a prospect strikes fear in most conventional ethicists, for whom a strong distinction between "natural law" and "positive law" has been a traditional cornerstone, even when the former is deemed cognitively or practically unattainable (cf. Deane-Drummond et al. 2003). At the same time, it reminds us of scientific creationism's intimate relationship with the mechanical world-view, which includes not only, say, marveling at God's machine-like handiwork in the construction of cells but, more importantly, inspiring humans to reverse engineer these divine machines to make them fit for distinctly human purposes, which God equally licenses.

2

Ideological Dimensions of the Problem

1. The Changing Ideological Temper of Our Times

To an old-school Marxist, the evolution–creation dispute taps into structural contradictions in the constitution of modern society merely "at the level of ideology," albeit an ideology that seems to travel increasingly well across national borders. While the Marxist may be right about the motives of particular people and groups who choose sides in this dispute, that still does not explain why a dispute about the nature of life remains such a compelling currency for transacting so many modern-day social conflicts. The answer here lies in the capacity for change that the two sides offer: creationism offers the hope for a radical transcendence of humanity's animal origins that evolution firmly denies. The shifting politics of this difference is noteworthy. Socialism and other ideologies of the left originated as late Enlightenment seculariza-tions of creationism, once humans replaced God as masters of the universe. Evolutionism's consistent complaint against this tradi-tion, in both its sacred and secular form, has centered on its dangerously unrealistic expectations, which fueled the great night-mare utopias of the 20th century, and arguably the climate change crises of the early 21st century. Interestingly, the decline of social-ism has shifted evolutionism decisively from the right to the left of the political spectrum, reflecting an acceptance of diminished expectations of what is humanly possible – or perhaps even that it is possible to be human (Singer 1999).

The conceptual battleground on which science and religion are most likely to face each other in the 21st century is the nature of *humanity*, namely, properties that all human beings possess either individually or collectively, but in any case uniquely as members of the same species. According to both social contract theorists like Hobbes, Locke, and Rousseau, and the classical sociologists Marx, Durkheim, and Weber, the full realization of these properties – e.g. language, government, economy, art, science – requires society. We may be each born with the poten-tial for these distinctly human qualities, but we need to be orga-nized in a certain range of ways to manifest them. However, the declining fortunes of the social sciences as a body of knowledge clearly demarcated from the humanities and, especially, from the natural sciences, in the last quarter of the 20th century, reflects an increasing uncertainty of what, if anything, is so special about being human (Fuller 2006b). Indeed, one very popular presenta-tion of evolutionary psychology, Steven Pinker's *The Blank Slate*, starts precisely by debunking the intellectual lineage of social science to which I have just alluded (Pinker 2002).

Intelligent design theory (IDT) joins the debate at this crucial juncture, reasserting the traditional Western idea that humans carry a spark of the divine lacking in all other living creatures. That we are a volatile mix of the divine and the animal may be traceable to theological mysteries surrounding the person of Jesus in Christianity. But it also survives in secular metaphysics as the mind–body problem and the attempt to reconcile free will and determinism. What is sometimes (especially in Europe) cast as the central problem of social theory, the relationship between "agency" and "structure," also tends to be discussed along similar lines (Fuller 1998a). While there is something clearly at stake in these debates, it is not perspicuously brought out in the terms in which they are normally conducted. Much better would be to pose the question as follows: *are humans defined in terms of where they came from or where they are going – the actual past or the potential future?* In terms of the above debates, "the actual past" captures what is common to "the animal," "body," "determinism," and "structure," while "the potential future" captures what unites "the divine," "mind," "free will," and "agency."

To define humanity in terms of an unresolved temporal problem usefully highlights a defining ideological polarity of the early 21st century. The extremes are epitomized by, on the one hand, the *animal rights* movement and, on the other, the project of *artificial intelligence*. Interestingly, both poles think of themselves as politically "progressive" and, perhaps even more interestingly, are driven by what each regards as cutting-edge science. On the one hand, such spin-offs of the neo-Darwinian synthesis as sociobiology, evolutionary psychology, and behavioral genetics are daily providing evidence for the preponderant overlap between human and other animal natures. According to the principal theorist of animal liberation, Peter Singer (1975), this evidence warrants a redistribution of sentiment across species so that "the greatest good for the greatest number" includes all sentient beings. On the other hand, improvements in prosthetic technologies designed to maintain and enhance human performance – and the public's growing acceptance of them – suggests that, contrary to the neo-Darwinists, what we value most in our "humanity" may not be natural at all. Indeed, one of the most articulate representatives of this position, Ray Kurzweil (1999), speaks of ours as an "age of spiritual machines," a phrase meant to suggest that we could design entities that end up superseding us in the qualities we value most.

Singer and Kurzweil represent opposite ends of a political spectrum, the center of which is still dominated by those – social scientists – concerned primarily with human beings as they are ordinarily understood: i.e. potential voters and consumers. Yet, seen from the *long durée* of intellectual history, it is remarkable that the center has held as long as it has. Before the ascendancy of the nation-state as the guarantor of "society" in the wake of the Protestant Reformation, political discourse veered between pagan skeptics who, like Singer, were preoccupied with minimizing physical pain in all earthly creatures and Gnostics who, like Kurzweil, wanted to hasten the end of history so that the Elect could enjoy their richly deserved spiritual reward (cf. Davis 1998). The legal innovation of the *universitas* that arose in medieval Christendom – enabling, at first, the creation of churches, monasteries, guilds, and universities, but later states and firms – historically held the center ground by promoting a non-biological mode of social reproduction that was legitimized by reference to the ultimate ends of the reproduced corporate entity. It constituted formal recognition of a strictly "immaterial" sense of survival, at least in that it did not essentially involve consideration of the members' family origins – only the beliefs and desires they themselves professed and would like to have realized (Fuller 2006b).

The weakening of the hold of the entities descended from the *universitas* on social life accounts for both the decline in social science's epistemological salience and the resurgence of a premodern ideological sensibility: the *left–right axis* now replaced by a renovated version of the orthogonal *sensate–idealistic axis*, to recall the terms introduced by the founding chair of Harvard's sociology department, Pitirim Sorokin (1970). The sensate axis is historically associated with the "therapeutic" strand in Hellenistic philosophy and the Eastern religions, the idea that humans should simply pass the time as painlessly as possible until death provides eternal release for the soul. The idealistic axis corresponds to Gnosticism as a movement familiar from the more zealous strains in Judaism and primitive Christianity, later revived as a source of civil wars during the Protestant Reformation. It called for a "revolution of the saints" whereby a spiritual vanguard would destroy all earthly institutions in order to hasten the Final Judgment (Voegelin 1968). The history of modern politics testifies to the difficulties in avoiding these extremes, with Freudianism and Marxism their respective exemplars, at least in popular culture.

I said that Singer and Kurzweil do not occupy quite the same social roles in their re-enactment of Sorokin's polarity. The former is not the therapist of classical skepticism and the latter not the firebrand mystagogue – what we might now call a "terrorist" – of early medieval Gnosticism. A better way of casting their difference is in terms of the chemical element that each takes to bring out the essence of humanity: Singer's *carbon versus* Kurzweil's *silicon*.

The first term of the binary is perhaps easier to grasp. The neo-Darwinian absorption of *Homo sapiens* into an undifferentiated gene pool reflects the carbon-based origins of all forms of life, which come to be differentiated into species through various compounded historical accidents (aka natural selection). This fact alone has been sufficient to motivate the range of field and lab studies dedicated to reducing the evidential difference between human and animal qualities, thereby lending increasing intuitive support for Singer's species-egalitarian ethics.

As for the second term of the binary, silicon is the element common to glass and other conductors of light and electricity. These materials have underwritten, on the one hand, the recent revolution in information and communication technologies that have enhanced humanity's interactive potential – or "interconnectivity" – and, on the other, the increasing acceptance of prosthetic extensions to the lives of individual humans, from implanted silicon chips to plastic surgery. Moreover, silicon's lure reaches into the remote past. I allude here to the historic fascination with *optics* as the interface science between God and his creatures, starting with Al-Kindi in the 9th century and eventuating in the dominance of visual metaphors for unmediated veridical knowledge in the modern philosophical imagination (Turbayne 1962).

As suggested above, excluded from this elemental binary is the artificial person, the incorporated form of social life rendered in English as "society" (in its institutional sense), or *Gesellschaft* in German, or *universitas* in the original Latin. Here classical social contract theory redeems its relevance to contemporary sociological discussions. All versions of the social contract presuppose that individuals come to see it in their own interest to combine in ways that force them to exchange one kind of freedom for another, at least putatively more valuable, kind. The maintenance of this contract entails the construction of technological means for organizing, monitoring, and, when necessary, disciplining the individuals – the overall result of which is (hopefully) a thriving social organism.

The most natural understanding of this process in terms of our binary is that carbon-based creatures employ silicon-based means to empower themselves in ways that in the long term become normative standards in terms of which the creatures themselves come to be evaluated and, where possible, improved. Thus, the passage of light through an undistorted lens became the original model for the frictionless medium of thought communicated from God to his creatures. This image was socialized in Jeremy Bentham's "panoptical" total institutions, as notoriously recounted by Michel Foucault (1979). The final stage occurred in the early 20th century with the discovery of superconductivity, which permitted the generalized high-speed conveyance of electromagnetic impulses – and hence the information potentially carried in them, as computers routinely do these days.

The "logic" of this historical trajectory is driven by increased *efficiency,* which, metaphysically speaking, consists in minimizing the matter needed to convey the same form. From that standpoint, the carbon end of the carbon/silicon binary appears "conservative" in its attachment to matter at the expense of form. Here lies the source of much contemporary "back to nature" environmentalism (aka biodiversity) that defines the *summum bonum* in terms of enabling the survival of the widest range of carbon-based creatures. In contrast, proponents of the "silicon" end of the binary aim to expedite the drive toward efficiency. This more "liberal" attitude toward matter simultaneously suggests a host of superficially unrelated ideological associations: Gnostic spirituality, revolutionary politics, artificial intelligence, and, of course, the kind of technological determinism that Marx both admired and feared in capitalism.

I say "superficially" because certain silicon-oriented individuals already embody this curious combination of sensibilities. A notable case is George Gilder, a founder of Seattle's Discovery Institute, the think-tank most openly dedicated to the promotion of IDT as an alternative to neo-Darwinism for a place in US high school science textbooks. This is probably the highest profile forum in which the battle between carbon- and silicon-based ideologies is fought today. Gilder, a trained economist, began life writing speeches for the liberal Republican Governor of New York, Nelson Rockefeller, and anathematizing the reactionary politics of Barry Goldwater. When Rockefeller's star faded, Gilder became an adviser to Richard Nixon, who managed to win two presidential elections (1968, 1972) by occupying the political center.

A sense of Gilder's political animus can be gained from *Wealth and Poverty* (Gilder 1981), a best-seller that argued for a renewal of "The Protestant Ethic and the Spirit of Capitalism," to recall the title of Max Weber's classic work – that one serves oneself best by serving others. Gilder championed supply-side "Reaganomics" as a policy to encourage such strategically self-sacrificing entrepreneurs. Here he capitalized on the atavistic strain of American Puritanism on which Ronald Reagan re-launched the Republican Party. At the same time, Gilder alienated followers of the philosophical novelist Ayn Rand, whose atheistic and egoistic brand of libertarianism was epitomized by Gordon Gecko's infamous slogan from the 1987 film, *Wall Street*: "Greed is good!" By the end of the 1980s, Gilder presaged the general trend toward the miniaturization of processes and products, which he called "quantum economics," but which nowadays is better known as "nanotechnology" (Gilder 1989). When not penning visionary opinion pieces for *Wired*, the fashionable on-line information technology magazine, Gilder sponsors conferences that bring together the likes of Kurzweil with intelligent design theorists (e.g. Richards 2002).

This sketch of the political implications of the carbon–silicon divide suggests a couple of points. First, the binary is genuinely "orthogonal" to the left–right divide in that it drives a wedge into the constituency of both the left and the right. Roughly speaking, the left is now divided between those who see Darwin and Marx as the beacon of progress, and the right is now divided between those who see the "natural" and the "artificial" as the primary source of value (Singer 1999; cf. Fuller 2006b: chs 4, 13). The former grounds the split between Greens and Reds, the latter the split between the traditionalists and the libertarians. The second point is that social science exists in a space that neither denigrates nor extols efficiency as a value – but seeks an "optimal" (as opposed to a "maximal") level of efficiency in terms of which the most individuals can live the most fulfilling lives. Here I allude to Bentham's classical definition of welfare as "the greatest good for the greatest number." Only entities satisfying that principle of incorporation are properly counted as "social," the product of which constitutes a *universitas*.

As it turns out, Bentham's maxim was coined by Joseph Priestley (Snyder 2006: 276–8). Priestley is nowadays known as the chemist who first experimentally isolated oxygen in the 1770s but misidentified it as "dephlogisticated air," leaving it to his correspondent and rival, Antoine Lavoisier, to give the gas both the name by which we know it today and its status as a chemical element. Indeed, Thomas Kuhn's mentor, Harvard President James Bryant Conant, himself a chemist, was instrumental in demoting Priestley's scientific contribution (Fuller 2000b: 181). This is not surprising, once Priestley's science is put in the context of his political views. While Conant saw science as a force for stabilizing society in volatile times, Priestley's position was anything but that. An ordained minister, Priestley spent most of his career defending Unitarian Christianity, a historically heretical view secretly shared by several leaders of the Scientific Revolution, not least Isaac Newton (Brooke 1991: 177–81). To be sure, the view was no less controversial in Priestley's day than in Newton's, despite the century that separated them. Priestley's Birmingham (UK) home was burned down and he spent his final years in Pennsylvania, albeit feted by such intellectual fellow-travelers as Benjamin Franklin and Thomas Jefferson (Commager 1978: ch. 2).

Dogmatic Christians of many denominations have regarded Unitarianism as among the most virulent of heresies because it treats Christianity's strong analogy between human and divine intelligence as a relation of converging identity. Nevertheless, Unitarian sensibilities motivated the idea that society can be constructed from first principles just as the deity constructed the universe. In this way, a natural-historical view of the human condition, epitomized by such blood-based forms of cultural transmission as royal dynasties, yielded to the modern contractarian view of society as a collectively beneficial artifice. Even more controversial, but increasingly relevant in our own times, is the idea that neo-Darwinism's natural-historical perspective might yield to IDT as the metatheory of life. Beyond strictly scientific questions of intelligent design's empirical status, traditional Christian fundamentalists have questioned the religious probity of humans presuming to understand the "irreducibly complex" units from which life is created. However, as we shall now see, this has been very much the *telos* of the intelligent design perspective, namely, that humans end up "playing God."

2. The Problematic Status of Humanity in Intelligent Design

The history of Western science can be modeled on the tale told by Leo Strauss of the history of Western politics. Straussian political historiography presupposes the promulgation of a "double truth," a dangerous esoteric one for the elite and a safe exoteric one for the masses. As Strauss (1952) originally observed, this doctrine is common to Jewish, Christian, and Muslim philosophical writing of the Middle Ages. The ultimate precedent was Plato's dialogues, which are simultaneously manuals to instruct the elite in maintaining social order and treatises on the gods to render the masses docile. According to Strauss, most philosophers have strategically blurred these two functions under the rubric of "metaphysics," leaving the impression that the mass account is superior to the elite one by virtue of causal priority. Thus, we are led to believe that the gods laid down an order that the elites simply have the privilege of understanding and administering.

Philosophers who failed to abide by this double coding of their texts have been (perhaps justifiably, in Strauss's view) subject to persecution, mainly by others but sometimes by themselves. In Strauss's own youth, this fate reached tragic dimensions in the case of Friedrich Nietzsche, who died insane after a promising start as a classicist. Strauss's diagnosis was reissued as popular prophecy by his student Allan Bloom, translator of Plato's *Republic*, whose 1987 best-seller, *The Closing of the American Mind*, is probably the definitive condemnation of the publicly expressed radicalism of the two generations of professors who were inspired by first Herbert Marcuse and then Michel Foucault.

A corresponding historiography of science in the Straussian style would focus on the difficulties in identifying Isaac Newton's "logic of discovery," the method he used to arrive at the most intellectually and practically powerful unification of human knowledge ever devised: is *Principia Mathematica* simply a human representation of the divine plan or a blueprint for completing a job half-done? Is Newton's ultimate purpose to revere and protect nature or to master and improve it?

To the casual (yet probably correct) observer, *Principia* was composed on a top-down basis, whereby Newton postulated three laws and a principle of universal gravitation, then managed to deduce the actual motions of bodies on Earth and in the heavens. The impression is reinforced by the book's Euclidean geometric style of presentation. However, this interpretation would suggest that Newton had successfully simulated the divine plan in his own mind, a potential sacrilege to his Anglican Christian audience, who, despite having split from Rome, thus reducing the clergy's need to mediate the human and the divine, nevertheless maintained an absolute difference between the human and the divine intellect. This explains the ignominious fate that befell the successor to Newton's chair, William Whiston, who insisted on developing and publicizing Newton's Unitarianism, even against the master's own explicit wishes (Fara 2002: ch. 3).

It is thus unsurprising that when discussing his method, Newton expressly declared, "*Hypotheses non fingo*" ("I do not feign hypotheses"). Newton wished to give the appearance that he had inferred the structure of the physical universe from the empirical phenomena – that is, from the standpoint of one of God's creatures, an imperfect recipient of the divine plan. By so shifting the epistemic grounds of his insights from deduction to induction,

Newton had removed the main theological obstacle to others pursuing his path of inquiry, namely, the very idea that one of God's creatures could successfully adopt the creator's point of view. Yet, I believe, this is precisely what Newton had done, at least much better than anyone up to that point in history.

Perhaps the clearest legacy of Newton's strategic ambiguity appears in the distinction between arguments *from* and *to* design in nature that were staples of natural theology in the nearly two centuries that divided Newton's *Principia* from Darwin's *Origin of Species*. A pivotal mediating figure was William Whewell, Master of Trinity College, Cambridge, and the leading natural theologian of the mid-19th century. He is nowadays known for having coined "scientist" to name a specialized profession, and "heuristics" to refer to mindsets that facilitate problem-solving. For Whewell, one biblical teaching stood out as a heuristic for science: *that humans are created in the image and likeness of God.* This claim can be taken in two ways. One stresses the similarities; the other the differences between humans and God. Are we junior creators or senior creatures? Junior creators reason from hypothetical causes to sensible effects, while senior creatures try to infer causes from effects. The former promises an argument *from* design; the latter *to* design. The former fuels ambition and possibly sacrilege, while the latter instills a humility verging on mystery.

This dual reading of the biblical claim captures the difference between the scientific attitudes of Isaac Newton and Charles Darwin. Newton presented his mathematical physics as the divine plan that was implicitly written into the Bible. He clearly thought he had got into God's mind. In contrast, Darwin pursued the humbler path of William Paley's analogy of nature's order to a watch found on the ground, implying the existence of a watchmaker. Unfortunately, the fossil record revealed to Darwin only a lot of broken half-watches, nothing that could have been produced by a God worthy of human respect. Darwin's humility remained but his faith disappeared. In today's secular culture, Darwin is more readily embraced than Newton as a scientific icon, though Newton remains unquestionably the greater scientist. Whereas Newton's life appears to imply that the Bible can provide a sure path to great science, Darwin's biography projects the politically correct image of a Christian who loses his faith through scientific inquiry.

However, it would be mistaken to suppose that enthusiasm for Newton vis-à-vis Darwin has split along theistic/non-theistic lines. On the contrary, both theists and non-theists have generally found Darwin's scientific personality much more palatable than Newton's. Even among the faithful, admiration for Newton never extended far beyond the undeniable brilliance of his achievement, whereas until the general anti-scientific backlash that followed World War I, Darwin's tentative manner made him appear very credible to theologians, perhaps even more so than to his fellow biologists, who became frustrated with Darwin's refusal to provide a clear theory of inheritance. For example, it was the founder of modern American fundamentalist theology, Benjamin Warfield, who, in the spirit of sympathetic understanding, first made the connection between Darwin's lifelong bouts of depression and his loss of faith. At the same time, Warfield proposed to reconcile Darwin and Christianity by a move that has become a staple in "theistic evolution," namely, that natural selection and the biblical narrative told the same story from the respective standpoints of efficient and final causes (Livingstone 1984: 116–21).

Newton's fan base consisted of those who believed that God was compelled to create humans in order to complete the divine plan: God may be the architect but humans are the builders. This explicitly anthropocentric take on cosmology inspired Charles Babbage, a contemporary of Whewell's at Cambridge and a successor to Newton as Lucasian Professor of Mathematics (the chair currently held by Stephen Hawking), to propose a remarkable account of humanity's relationship to God, based on recent innovations in probability theory. According to Babbage, God programed the universe with stochastic variables, the values of which are supplied by free human acts, thereby rendering God the source of all power without exerting total control. (The language of "programing" is entirely appropriate to Babbage, whose plans to build an "analytical engine" are normally regarded as the first clear formulation of the modern digital computer.) The result is a kind of "principal-agent" theory of creation, in the jargon of economists (where God is the principal and humans the agents), or what John Stuart Mill sympathetically dubbed the "limited liability God" (cf. Knight 2004: 49–51).

Thus, Newton's public relations problems with the devout went beyond his well-documented misanthropy, egotism, and paranoia. That one could explain and predict all physical motion by conceptualizing the universe as a big machine exceeded orthodox theological – but, of course, not heretical – expectations of humanity's ability to approximate the divine intellect, or "intelligent designer," as we say today. Newton enabled the heretics to join the mainstream, claiming that humanity had matured sufficiently to take over the reins of creation from God. What had been Gnostic fantasies of expediting the course of history before Newton became secularized by the likes of Condorcet, Comte, Hegel, and Marx, from roughly 1770 to 1870, as theories of social progress (Passmore 1970: ch. 10).

Today, mechanistic approaches to nature as prescribed by Newton are criticized for oversimplifying complex organic processes. Yet it is worth stressing that the mechanical world-view was originally subversive for suggesting that God is just a big mechanic, and not some opaque animating force. For mechanists, biology reduces to a kind of divine technology. Animals – including humans – are products of artifice in exactly the same sense as our own machines. Indeed, mechanists have expected that technological progress would enable us to fathom life's mysteries by stimulating our capacity to mimic God's handiwork (Noble 1997). Thus, Newton read the Bible as a personal intellectual challenge to "reverse engineer" the divine plan. A century later, Kant was so impressed by Newton's self-understanding that he argued that the full realization of our humanity depends less on God's existence than our assumption that God exists. This possibly profound (Kant called it "transcendental") truth about human psychology lay behind Whewell's views about the sources of scientific creativity.

In updating the mechanical world-view, contemporary IDT is less a rival theory of life to Darwin's than a more ambitious "science of design" that is indifferent to the life/non-life distinction. This shift in scientific focus helps to explain theory's peculiar mode of reasoning – why, say, in the most popular book on intelligent design, the biochemist Michael Behe (1996) moves so easily between reasoning about the design of mousetraps and cells. Not surprisingly, chemists, engineers, and computer scientists – rather than those who study nature in the wild – have been most attracted to this perspective. Indeed, the best evidence for "atheism" among biologists is to be found among those closest to Darwin's natural history expertise. In contrast, evidence for accommodation, and perhaps even endorsement, of IDT can be found, say, in the guarded pronouncements about evolution made by the director of the Human Genome Project at the US National Institutes of Health, Francis Collins, a born-again Christian. He claims to have decoded "the language of God," a phrase echoed by President Bill Clinton in 2000, upon completion of Project (cf. Collins 2006). Collins described the completed genomic map as a "shop manual with a detailed blueprint for building a human cell." Whenever asked to comment on "evolution," Collins typically restricts the theory to a demonstration of the interrelatedness of all living things through a line of common descent.

Collins's engineering imagery is not accidental but points to the main constituency for IDT among scientists, namely, those who think of themselves as doing on a smaller scale (or perhaps bringing to completion) work that the creator has done on a grand scale. Thus, the leading scientific proponent of IDT in the UK, Andrew McIntosh, Professor of Thermodynamics at Leeds University, participates in a major state-funded (via the Engineering and Physical Sciences Research Council) project on "biomimetics," an emerging field of biotechnology that treats organisms as prototypes for humanly useful things and processes. Historical examples include the inspiration that birds provided for aviation, insects for the military, and fruits for the packaging and transport of produce (on the origins of this field from early 20th-century biophysics and mid-20th-century bionics, see Rosen 1999: chs 18–19). While support for creationism or IDT is not necessary to engage in this work, it would be difficult to ground the plausibility of such research in the first place without seeing animal and plants as designed to perform functions that might inspire humans to improve upon them through technology. The biblical roots of this idea are clear, but it has not been the attitude that the bulk of humanity has held toward nature for most of history. The default attitude has more resembled today's Darwin-backed ecologists who regard other organisms as co-habitants with humans and therefore entitled to be treated not as mere means but as ends in themselves.

Because intelligent design theorists have generally not hidden their theological motives, it is easy to forget the unabashedly secular precedents for their view, especially the "sciences of the artificial," to quote the title of Herbert Simon's (1977) visionary proposal to make instrumental rationality the overarching principle for the organization of all scientific inquiry (cf. Fuller 1985). For Simon, "design" is about the realization of ends through the available means, which are always presumed to be sub-optimal (i.e. incurring costs, forcing trade-offs). This general idea is common to all problem-solving activities, be they performed by humans, animals, machines – or, for that matter, the God of the Old Testament, who must realize even his ideal creature in a resistant material medium.

The feature of Simon's proposal that has probably received the most philosophical attention, yet without much historical or sociological scrutiny, is his account of what he calls "the evolution of complex systems" (Simon 1977: 200–29). Here Simon shows how an intelligent designer could simulate the pattern of evolution associated with the neo-Darwinian synthesis. Simon begins by noting a problem that information theorists posed to evolutionists as early as 1955 – that is, about a decade after the synthesis was finally forged between molecular genetics and natural history. If organisms evolve through the kind of chance variation and selective retention stipulated in the synthesis, then the four billion year timeframe implied by the radiometric dating of fossils would be insufficient to produce the biodiversity we see today, including the emergence of humans.

Simon's solution appears in the parable of two watchmakers, Tempus and Hora (respectively, the Latin and Greek words for "time"), itself a nod to Paley's original presentation of the argument for divine design. Simon asks the reader to imagine evolution as an extended project where interruptions are to be expected. The process is thus neither blind nor completely determined at the outset. Construction proceeds not by trying to assemble the ideal object directly, but by a series of sub-assemblies that remain intact even in the face of interruptions, so that they can be built upon later.

Taken literally, this image suggests a God who arrives at a satisfying solution through trial and error, very much like a technologist. All prior organisms are effectively rough drafts or prototypes that are retained until a better version is designed. From a biological standpoint, this recalls Ernst Haeckel's perceived parallelism between embryology and evolution: "Ontogeny recapitulates phylogeny," that is, an embryo undergoes a metamorphosis that appears to repeat earlier evolutionary stages. In theological terms, God may have the power to realize his plan ultimately, but he may not know exactly how until he has actually applied his ideas to matter and observed the consequences. That the God of Genesis does not create all at once, but requires six "days," already suggests as much.

Scientific descendants for this view begin with the "catastrophist" accounts of natural history associated with Georges Cuvier and Louis Agassiz in the 19th century, which interpreted the stratified character of the fossil record (i.e. the tendency for ecologies to be replaced *en masse* with significant climate change) as evidence for the Noachian flood. While Cuvier and Agassiz studiously avoided commenting on the cognitive limitations of the deity that might have necessitated this piecemeal approach to creation, a much bolder line was taken by Robert Chambers in his cause célèbre, *Vestiges of the Natural History of Creation*, a bestseller in 1844. Chambers scandalized Victorian readers – and alerted the young Charles Darwin to the potential controversy surrounding his own views – by arguing that science reveals God's mystery to lie simply in his inelegant *modus operandi*, whereby destruction is part of his creative process (Secord 2001). The most recent and fully secularized version of this position is the theory of "punctuated equilibrium" advanced in the 1970s by Niles Eldredge and Stephen Jay Gould.

Setting aside the deity's possible deficiencies, there is little controversy that humanity's capacity to design living things evolved slowly at first from the selective breeding of animals and plants, first depicted in Egyptian hieroglyphics around 4000 BC, which, as Richard Dawkins likes to observe, is when biblical literalists typically believe creation occurred. Our ability to design life began to pick up the pace with the biochemical synthesis of organic materials in the 19th century, before accelerating to the full-fledged sciences of genetics and molecular biology in the 20th century.

It is only once we reach these latter stages that Darwin's own theory of natural selection can be said to have acquired a proper sense of "mechanism." I mean here the role of Mendelian genetics in forging the neo-Darwinian synthesis in the 1930s (Ceccarelli 2001; Smocovitis 1996). Contrary to popular accounts, Darwin failed to offer a properly mechanistic account of evolution because he lacked a credible theory of hereditary transmission. Indeed, by the time *Origin of Species* reached its fiftieth birthday in 1909, Darwin's theory was itself close to extinction in biology – though it continued to enjoy admirers in theology and sociology (Bowler 1988). Darwinists could offer only "just so" adaptation stories to explain species change. However, their theory was saved by the rediscovery of the work of Gregor Mendel, a Catholic monk whose statistical account of heredity was designed to capture the range of traits that God deemed permissible in a given species. Mendel was no evolutionist, but a special creationist with a grasp of probability theory.

Mendel's three laws of inheritance can be interpreted as design features of the divine plan because they can be specified without reference to the environment in which species reproduce. In short, the laws appear as if they were programed into each generation of organisms. The three laws, now captured in a quadratic equation (the Hardy-Weinberg Principle), are as follows: the Law of Independent Segregation states that traits are regularly reproduced across a population, not blended over generations; the Law of Dominance states that traits are not equally distributed in a population; and the Law of Independent Assortment states that traits are not inherited as a package. Epistemologically, the laws look like lessons inferred by a gambler who has kept track of his bets against the house and, after a period of trial and error, managed to succeed. The gambling house in this case is run by

God, and "success" means the development of new breeds, and perhaps even species of plants and animals, not included in the original divine plan but which continue to reproduce. Two opposing religious responses to such success are possible. One is to anathematize those who claim success, as happened to magicians and alchemists until physics and chemistry became respectable scientific subjects in the modern era. A milder version of such ostracism may explain why religious authorities allowed Mendel's original papers to languish once scientific experts rejected them. In the end, Mendel had to resort to a newly established "Natural Sciences Society" in his native Moravia, though even this society's members failed to see the far-reaching practical consequences of his statistically driven theories (Wood and Orel 2005). This fate is akin to the standard response of casinos. Players who become good enough to beat the house at blackjack (or "twenty-one") consistently are barred from further play.

However, a second and more reasonable religious response is that, by extending human dominion over nature via agricultural innovations, Mendel's laws simply contributed to the completion of the divine plan, as might be expected eventually to happen, if we are indeed created in the image and likeness of God. Moreover, once the history of genetics is treated as distinct from that of neo-Darwinism, the "playing God" theme becomes even clearer over the 20th century, especially with advances in molecular biology and now biotechnology. Geneticists have been always keen to speed up the processes of evolution by experimenting on species like the fruit fly whose reproductive patterns might allow billions of years of change to occur over, say, the six days stipulated in Genesis.

The sense of "design" relevant to Mendelian genetics is captured by the leading contemporary intelligent design theorist, William Dembski (1998), who seeks a sense of constraint in nature that is not reducible to necessity, chance, or some combination of the two. Dembski, a trained mathematician, was inspired by the apparent impossibility of designing a random number generator (Dembski 2004: 311–13). For any seemingly random series of numbers, it is ultimately possible to infer the program that generated them. This throws into doubt the reality of pure chance without, at the same time, vindicating necessity. Dembski wishes to claim the middle realm between chance and necessity for "design," which he calls "complex specified information."

There are at least three secular precedents for seeking such an ontological "third way" for design. The first comes from Charles Sanders Peirce, who tried hardest to convert philosophical pragmatism into a full-fledged metaphysical doctrine. Peirce coined the term *agapism*, alluding to the kind of love the New Testament God has toward his creatures, to capture the sense of design that emerges over the evolution of the universe but is not reducible to chance and/or necessity. Despite the biblical overtones of his own formulation, Peirce was getting at a generalized sense of sociality, one pertinent to the organization of both organic cells and human societies, common to contemporaries like Herbert Spencer and Gabriel Tarde.

A second precedent, perhaps the one closest to Dembski's own concerns, comes from the mathematical version of information theory associated with Claude Shannon and Warren Weaver. Accordingly, if a receiver deems a signal neither redundant (i.e. predictable) nor noise (i.e. accidental), then it is informative (i.e. meaningful). Thus, Dembski's "complex specified information" looks like a version of what the original cyborg anthropologist, Gregory Bateson (1979), dubbed "the difference that makes a difference." The key point here is that information is always relative to the receiver, which would support the biblical view that reality has been created so that we can make sense of it. In that case, the genetic code should be understood as *literally* informative – that is, a message inscribed for us to decipher.

A final and perhaps ironic precedent can be found in the strategy adopted by socio-biologists and evolutionary psychologists *against* social scientists to explain the cross-cultural range of human behavior: People do not behave exactly the same regardless of environment (i.e. not pure necessity), but neither does their behavior simply mimic their environments (i.e. not pure contingency). This allows for a more nuanced sense of biological determination, a pre-programed range of variability that accounts for limits in the human capacity for change. Little surprise, perhaps, that the intellectual godfather of this extension of neo-Darwinism, the Harvard entomologist E. O. Wilson (1998), has recently drawn on his Alabama Baptist roots to persuade the religious to channel their design-oriented impulses to an evolutionary cause (Wilson 2006). Sociology's loss could thus become theology's gain, were IDT to join forces with genetics to provide a foundation for a new design-based social science. A work that might be consulted for such a project is Milbank (1990).

3. Conclusion: Science as the Ultimate Means of Human Transcendence

In the two centuries that separate Francis Bacon from William Whewell – the one the founder of the "scientific method" and the other the coiner of the word "scientist" – science was distinguished from philosophy and theology by its appeal to something called "induction." But unlike current understandings of this term, "induction" did not mean for these thinkers the simple registration of events, as a passive understanding of the *tabula rasa* metaphor so favored by empiricists would suggest. Rather, induction involved the re-specification of competing hypotheses in terms that could be subject to a decision by a judge who would set up a test neutral to the competing sides. Bacon called this test the *experimentum crucis*, the "crucial experiment."

Bacon's position as Lord Chancellor (i.e. the King's lawyer) just before the religiously inspired English Civil War is significant. Once Henry VIII broke with the Church of Rome in 1532, Britain had become a religious free zone. Bacon presumed that claimants to the sort of knowledge that warrants the title "science" would wish to claim privileged access to the mind of God. However, to avert what his secretary Thomas Hobbes would later call "the war of all against all," Bacon wanted to force all such claimants to reword their claims from the standpoint of a humble creature who knows "bottom-up," so that the judge, as God's proxy, might decide between them.

Three points about this strategy are worth observing.

1. Bacon was influenced by the inquisitorial mode of justice preferred on the European continent, in which the judge establishes the field of play between competing theories by compelling their advocates to pronounce on an unresolved empirical case by an agreed procedure (Franklin 2001: 217–19). The "track record" of the competing theories – the currency in which inductive knowledge claims are transacted – is established in this setting, not by the sort of common-sense anecdotalism that often passes for "induction" these days (e.g. the reliability of the sun rising every morning). In this respect, what I have called the "Baconian fantasy" would be ideally expressed as a regularly published *Consumer Reports* that "test-drives" competing knowledge claims along a variety of dimensions.

2. Bacon was essentially proposing a negotiated exchange of power that presaged Hobbes' more drastic presentation of the social contract, 30 years later, in *Leviathan*. The epistemological equivalent of "laying down arms" is the conversion of deductive to inductive knowledge claims. The Baconian state, as God's executor on Earth, would assume responsibility for setting down the (deductive) premises in terms of which claimants to knowledge must then make their (inductive) case. Put less abstractly, the state's power would be acknowledged by the knowledge claimants reformulating their claims in a common empirical currency. In this context, the compulsory presentation of evidence (in exchange for credibility) is formally expressed as voluntary submission to a superior force. These origins persist in the phrase, "the force of evidence," to describe objective grounds for accepting a knowledge claim.

3. Bacon's interest in promoting a "scientific method" was fundamentally different from many who later claimed him as a precursor, not least John Dewey (1953: ch. 2). Bacon deployed the scientific method as a forensic device for distinguishing policy-relevant information from religiously inspired metaphysics, whereas Dewey wanted to teach children the scientific method to counteract any such home-based metaphysics. The difference between the lawyer's and

the educator's perspective here is subtle but significant. Bacon accepted the inevitability of religious divisions and fractious politics but refused to let them interfere with statecraft. In contrast, Dewey saw scientific training as ultimately enabling Americans to abandon religion's destructive tendencies, the very ones their European forefathers tried to escape without complete success.

Bacon, like his protégé Hobbes, saw the state's need for collective intelligence as a bulwark against its relatively limited capacity to alter the human mind's tendency toward metaphysical enthusiasms. In contrast, Dewey was much more hopeful about the prospects for a fundamental change in human thinking to make it more problem-oriented and less dedicated to fixed ideas. The passage of 300 years accounts for much of this difference in attitude. Publicly supported secular education was becoming a reality in Dewey's day, whereas Bacon and Hobbes could only envisage the perpetual reproduction of ideological differences in religiously based schools.

As it turns out, Bacon became entangled in royal intrigues and his proposals for a scientific judiciary – "The Great Instauration" – were never institutionalized, and England subsequently suffered a civil war (Webster 1975). However, shortly after the war, the Royal Society was founded to establish Bacon's principles on a small scale that enabled science to remain independent from but not hostile to the reinstated monarchy. In this context, the Baconian judge's role was replaced by that of the collective opinion of the Royal Society's members, who (ironically) rejected Hobbes' candidacy for membership, even though his political philosophy captured the spirit of Bacon's original vision of the knowledge society (Lynch 2001).

This unrealized ideal consisted in the state, armed with science, adopting the role previously occupied by God, armed by the Church, as the arbiter of knowledge-power claims. The Hobbesian strategy was reinvented as *positivism* by Auguste Comte in the early 19th century and remained a century later in the logical positivist attraction to constitutionalism and even the idea of a neutral scientific language. Here the relevant witnesses are Hans

Kelsen and Otto Neurath, who embodied, respectively, liberal and socialist versions of this ambition.

However, positivism was always plagued with the problem of finding an adequate political vehicle for its aspirations (Fuller 2006a: ch. 4). By the final quarter of the 19th century, science had managed to unite people within nation-states while dividing states from one another in competing imperial ambitions and sublimated international conflict – sublimated, that is, until World War I. This period of contained violence rendered positivists like Ernst Mach better defined politically in terms of what they opposed rather than what they supported, which in practice often made for an ineffectual pacifism, a regular object of Marxist-Leninist suspicion and derision. Not surprisingly, with the passage of two world wars, the positivists abandoned the nation-state for the idea of world government. Thus, Kelsen was active in refashioning the Covenant of the League of Nations to make for an effective Charter of the United Nations.

Probably the clearest and most hopeful statement of the positivist vision of world government was UNESCO's founding document, authored by the agency's first director-general, T. H. Huxley's grandson, Julian (Huxley 1947). Julian Huxley's updating of Comtean themes makes for arresting reading, even today. Following Bacon, Hobbes, and the Enlightenment more generally, Comte had appealed to the need for *Homo sapiens* to abandon internecine conflict and face common foes that threaten to keep us mired in animality, holding us back from realizing our full "humanity," which Huxley regarded as a semi-transcendent state that is promoted nowadays under one of his coinages, "transhumanism" (about which more in chapter 5).

Comte had cast the common foes facing humanity as theologically and metaphysically fueled abstractions that created mythical divisions that can be overcome only by a scientific reorganization of *Homo sapiens*. Comte understood this process as involving a higher-order version of the reorganization of matter required for the constitution of living things. Writing over 100 years later, Huxley proposed much the same strategy for tackling a different set of foes, poverty and disease. UNESCO's founding document set the pattern for thoughtful responses to the world's systemic problems, especially by development agencies. However, interestingly – and perhaps tactfully – these documents (e.g. Sachs 2005) do not highlight, as Huxley did, the role that broadly eugenic policies might play in the rational reorganization of the world order.

Although Comte proposed an intellectual diagnosis and Huxley a material one, they agreed over what would constitute *failure* for the emerging social organism they called "humanity": *death*. Thus, progress toward the realization of humanity is measured in terms of the number of people who can live "freely" for as long as possible – be it free from censorship, oppression, poverty, or illness. This operationalization is most familiar as the utilitarian definition of welfare as "the greatest good for the greatest number." As I observed earlier, this definition, associated with Jeremy Bentham, was actually coined by the radical scientist-cleric Joseph Priestley seven decades before Darwin's *Origin of Species* provided a persuasive popular defense for the naturalness of not only individual death but also species extinction. Let us conclude by briefly revisiting Comte's and Priestley's "death-defying" scientism, ending with its surprising theological roots.

Comte's interest in the deferment and possible transcendence of death was a product of his medical studies at the University of Montpellier, where he learned of Xavier Bichat's then revolutionary call for physicians to abandon the ancient view of death as the final phase of an individual's life cycle in favor of regarding it as the ultimate challenge that nature poses to the efficacy of medical science (Albury 1993). However, Comte believed that Bichat still retained too much of the Epicurean fatalism associated with materialism, which led him to believe that the fight against nature was bound to be futile. In this optimism, Comte was influenced by Jean-Baptiste Lamarck, whose conception of evolution implied that organisms became increasingly competent over successive generations to alter the environments in which they must gain selective advantage for purposes of survival (Pickering 1993: 588–91).

Priestley, the great luminary of Unitarian Christianity, enters the picture as part of the hidden religious backdrop to this discussion that provides a link that extends back from Newton and forward to Huxley. Priestley was notorious among theologians for denying the immortality of the soul while affirming the resurrection of the body (Knight 2004: 9–11). Modern Christians uphold both doctrines, typically regarding immortality as the more philosophically defensible one. However, Priestley followed Newton in treating the immortality of the soul as a pagan atavism, traceable to Plato and filtered through the Gospel of John the Evangelist, that ascribed diminished ontological significance to matter, not least the human body. For Priestley and Newton, this was tantamount to denigrating the person of Jesus. To be sure, Newton and Priestley never disputed the ontological significance of the spirit. Rather, they defended a version of materialism that they believed was necessary for humans to make the most of their earthly endeavors. This version of materialism aimed to steer a middle course between the ideological extremes of pagan skepticism and Gnostic spiritualism that we saw in evidence at the start of this chapter.

In secular terms, at stake in the stress on bodily resurrection over the soul's immortality is the role of *memory* in the constitution of humanity. Whereas the *immortalist* would purge all reminders of earthly error and failure, the *resurrectionist* would preserve them as constitutive of our becoming who we are. Among the main legatees of the latter sensibility were such reflective historicists as Comte and Hegel, not to mention that progressive evolutionist Lamarck, all of whom retained a conception of "genetic memory" that persists today in ideas as diverse as the unconscious and culture itself.

3

Complexity as a Conceptual Battleground

1. The Problem of Defining Complexity

Intelligent design theory differs most markedly from other versions of creationism by the emphasis it places on *complexity*. The two leading intelligent design theorists, Michael Behe (1996) and William Dembski (1998), define the presence of intelligent design with reference to, respectively, "irreducible complexity" and "complex specified information." As we saw in the previous chapter, such complexity should be understood in a communicative context. This context consists of a sender and a receiver of information, where "information" is defined as something enabling the receiver to act in a way for which there would otherwise not be license: Bateson's (1979) "difference that makes a difference." To increase information, in this sense, is to resolve the uncertainty of the receiver, who can then contribute order to a given situation.

For the intelligent design theorist, the sender is clearly a deity who knows how to communicate with humans, perhaps including what they need to do to complete the divine plan. In any case, under the circumstances, complexity is *not* a brute fact of reality but an unintended consequence of our relationship to reality. Thus, complexity is measurable only relative to a frame of reference. This point was put well by the late Canadian biophysicist Robert Rosen (1999: esp. chs 18–19), who defined complexity by how the modeled exceeds the expectations of the modeler, thereby providing a source of surprise, or new information. It follows that a world that met all our expectations would appear very simple – as it might to God.

As a philosophically sophisticated scientist of the old (positivist) school, Rosen drew some suggestive conclusions from this observation related to the failure of models to capture their target realities: failures of syntax generate the need for a separate realm of semantics; failures of epistemology generate the need for a separate realm of ontology. The relationship between the two terms in each case is asymmetrical: whereas syntax or epistemology simply presents reality to a speaker or knower, semantics or ontology performs that function but also represents some other (typically earlier) presentation of reality (typically in terms of what is true or false).

Consider the relationship between God and humans vis-à-vis reality according to scientific creationists. The Bible suggests that God brings things into existence simply by naming them, very much in the spirit of world-building associated with subjective idealism or social constructivism. A strong Neoplatonic strain in Jewish, Christian, and Muslim philosophy characterizes God's perfection by saying that whatever God thinks must come into existence, since thinking per se is an unrealized state of being. Thus, whatever God says simply exists, and whatever he knows simply is. Humans, of course, are in a more difficult situation. Created in the "image and likeness" of God, humans also have their divine spark. It enables us to comprehend reality as well as we do through theories and models. However, our words and images are not self-validating. To think they might be is to risk the cognitive equivalent of idolatry, as prohibited by the Second Commandment. We are often wrong because reality exceeds (is more complex than) how it appears to us. Accounts of that difference result in ontologies and semantics that try to translate our simple understanding of reality into something approximating its true complexity. Theology moves away from religion and toward science – very much in the way that we saw Newton, Priestley, and Whewell did in the previous chapter – as one comes to think that the gap between the divine and the human epistemic positions can be closed.

These two levels of analysis – how reality appears before and after complexity is registered – have not been always kept straight, even in that most philosophical of languages, German. However, by the end of the 19th century, the physicist Heinrich Hertz, generally known as the discoverer of radio waves, had begun to stabilize the difference in meanings between *Vorstellung* and *Darstellung*: sensory presentation versus graphic representation; feeling versus deliberation; implicit model versus explicit model. Janik and Toulmin (1973) mark Hertz as an important transitional figure between Schopenhauer and the early Wittgenstein. On the one hand, Schopenhauer "sensationalized" Kant's domestication of reality through the categories of understanding by portraying this feat as an uphill struggle of the will against an onslaught of experience that eventually overwhelms us in death. On the other hand, the logical positivists read Wittgenstein's *Tractatus* as defending conventionalism as, so to speak, a triumph of the will over an otherwise unruly reality. (This imagery remains in the idea that a conceptual framework is "imposed" on reality.) The positivists were still left with the question of whether the phenomena that escaped the conventions imposed by scientific models were to be treated as cause for revision or retrenchment of the models. Karl Popper's Ph.D. supervisor, Karl Bühler, and Sigmund Freud exemplified the radical alternatives on hand in Viennese psychology before the rise of Nazism (Hacohen 2000: ch. 4).

This brief intellectual history should not obscure Rosen's original point: the refusal of the modeled to conform to the modeler's expectations operates as a reflexive check on the modeler, the result of which is the generation of a "second-order" discourse, or "metalanguage." The metalanguage then provides a systematic translation of things expressed in terms of the original model (i.e. the "first-order" discourse) to distinguish what does and does not conform to properties of the modeled. As Hertz might say, the *Darstellung* (metalanguage) is a reflection upon the limits of the *Vorstellung* (language). The legacy of this idea remains very strong in analytic philosophy, through the work of Alfred Tarski and Donald Davidson. It persists as a gloss on the correspondence theory of truth, where "truth" is the name given to the more complex reality revealed by the inadequacy of our models.

The metalanguage's translation function raises a feature of complexity that distinguishes it from mere *complication*. The great game theorist John von Neumann once said that a set of phenomena is complex if it is simpler to describe the structure that generated the phenomena than to describe the phenomena themselves. This structural description, captured in the metalanguage, would demonstrate the phenomena as the interactive product of a minimum set of principles, whereas a state description of the phenomena would merely list them as independent events. The latter would be complicated by virtue of each event being treated separately, but it would not be complex because the events are not systematically related to each other.

Put in more anthropocentric terms, we can distinguish complexity from complication in terms of relative *manageability*. Complex phenomena are manageable because they are reducible to a simpler account, whereby seemingly disparate events are successfully treated in a similar fashion. The profundity of this point would be difficult to overestimate. Whatever successes that planned economies have enjoyed, whether it be Oskar Lange's market socialism or John Maynard Keynes's socialized markets, have rested on the idea that anything desirable that has resulted from rather haphazard, self-organized means could have been reached more efficiently with greater foresight. (Here Keynes may have been inspired by Newton, whose theological papers he was among the first to see.) In contrast, one might say that the *raison d'être* of Ludwig von Mises, Friedrich Hayek, and the Austrian school of economics is the belief that no clear distinction can be drawn in practice between what we are calling "complexity" and "complication" (Steele 1986).

Complexity would thus seem to be identified with science, and complication with narrative or perhaps historical knowledge (as "one damned thing after another"). Perhaps the philosopher of the recent past who took this distinction most seriously was Imre Lakatos (1981), whose self-appointed task was to "rationally reconstruct" the history of science to show how the course of inquiry could have been expedited, had scientists adhered to his methodological precepts.

The diametrically opposite view – that complication is itself a desirable outcome of scientific inquiry – is exemplified in Bruno Latour's (1988a) self-styled "irreductionist" methodology for the study of technoscientific networks. For Latour, who only wants to follow the actors as they extend their networks, any resistance met along the way is an opportunity to reveal a new actor whose interests need to be negotiated for the network to continue. Latour always treats the problems and solutions to these situations as local, with all actors operating on a level playing field. There is no progress or error in network formation, only extension or termination. More importantly, there is no reflexive turn, which might cause a network to evolve into a hierarchy whereby the network itself comes to be managed as a unified object (Latour 1988b). The difference between Austrian economists and Latour's actor-network theorists is that whereas the former would simply fire the Keynesian central banker for lacking an object – "the economy as such" – over which to preside, the latter would allow the banker to function as one among many agents in the economic field.

2. Kuhn and the Complexity of Science

The complexity–complication distinction is relevant when understanding science as a historical process. Thomas Kuhn (1970) portrayed scientists laboring in a mature paradigm as engaged in puzzle-solving, which inevitably generates "anomalies," that is, puzzles that remain unsolved by the methods of normal science. These anomalies may come to be associated with ad hoc solutions that introduce arbitrary assumptions to supplement the paradigm. In any case, they are complications, which over time accumulate, eventually precipitating a crisis and shift in paradigm. The new paradigm then provides the metalanguage into which the terms of the old paradigm are translated. Thus, complication in the old paradigm becomes either a simple or complex statement in the new one. For example, before Lavoisier's Chemical Revolution, the element of oxygen was known as "dephlogisticated air."

That the sequence of paradigms in the history of science should be marked by successive degrees of complexity was not lost on logicians acquainted with Kuhn's theory. Although Kuhn is fairly portrayed as downplaying the role of logic in scientific theory choice, that did not prevent his own theory from being portrayed in logical terms. Thus, a postwar German follower of the logical positivist Rudolf Carnap, Wolfgang Stegmüller (1976), and a physics-trained US colleague, Joseph Sneed, spent the 1970s showing that scientific revolutions mark a rise in the level of complexity captured by a domain of inquiry. While Kuhn and fellow-travelers like Paul Feyerabend stressed the incommensurability of the fundamental assumptions of the old and the new paradigm, Stegmüller and Sneed highlighted the new paradigm's increase in scope, which comes about through a reorganization of the phenomena captured by the old paradigm into new categories, which in addition then allows new phenomena to come into view (Balzer, Moulines, and Sneed 1987).

However, this "model-theoretic" approach to science ended up having limited appeal. While it precisely represented the logical relationship between textbook presentations of scientific theories in the old and the new paradigm, it said nothing about how the former metamorphoses into the latter. Stegmüller and Sneed thus provided a retrospective rationalization for scientific complexification, the mechanisms behind which remained obscured. In response to this inadequacy, I would propose that the greater complexity associated with scientific progress relies on two factors: the obvious factor is the greater comprehensiveness of the new paradigm – Einstein's theory constitutes an advance over Newton's because it includes the latter as a special case; the subtle factor is what I would call *the redistribution of modal value*. In other words, statements (or events) that had appeared impossible or merely contingent now become possible or perhaps even necessary – and vice versa. This factor provides the fine structure of the "gestalt switch" associated with scientific revolutions.

Perhaps Kuhn's most profound but underappreciated insight was that scientific revolutions reconfigure a domain of inquiry in two senses at once: as both a legitimate heir to the old paradigm and a progressive research program in its own right. These two reconfigurations – one past- and the other future-oriented – are encoded in the new paradigm, which operates as a metalanguage for making sense of the old paradigm, rendering it in equal measures wrong and irrelevant. When Kuhn notoriously claimed that "the world changes" with each paradigm-shift, he meant just this: not only does the world now look radically different, *but also it looks like the world we should have been inhabiting all along*. This latter reflexive point, which involves the redistribution of modal value, is what leaves us with the impression that the new paradigm clarifies, rather than merely changes, the subject of scientific discourse. Kuhn rightly likened this subtle historical revisionism to the workings of the Ministry of Truth in Orwell's *1984*.

Thus, seen in the aftermath of the Chemical Revolution, Joseph Priestley is not credited with having experimentally isolated oxygen in its pure state, even though that is how his achievement would have looked to an observer from our time. This is because Priestley failed to see that he had discovered a physical element. Rather, he thought he had manufactured "dephlogisticated air." To be sure, Priestley recognized that this substance possesses oxygen's life-enhancing properties. But he thought he had done this by removing phlogiston's flammable quality. Thus, he saw his feat more in the spirit of a technological innovation than a fundamental scientific discovery. What Priestley regarded as a contingent application, Lavoisier treated as a necessary principle. Today Priestley is portrayed as a historical curiosity for having gotten "so close yet so far." But one could also say that Priestley would have taken chemistry down a radically different path, which Lavoisier closed off. Thus, Lavoisier did not merely correct Priestley's error but relegated Priestley's scientific vision to an impossibility predicated on the conceptually incoherent fire element, "phlogiston."

We see that Kuhn's model of scientific change, in somewhat different senses, both *registers* and *reduces* the complexity of the history of science. On the one hand, complexity is registered in that each successor paradigm generates a metalanguage that comprehensively redescribes its predecessor. On the other hand, complexity is reduced insofar as that metalanguage consists of statements that in translation redistribute the truth-values of statements made in the earlier paradigm so as to render its categories redundant. Thus, everything Priestley said that remains true is recoded in the language of oxygen, while his own language of phlogiston is literally rendered unspeakable (Kitcher 1978). Expressed in these terms, it becomes clear why Priestley is not a fixture in contemporary histories of chemistry: Anything he said that was worth saying has been said better by others. At least, that is the story officially told by the new paradigm vis-à-vis what came before it.

Suppose we accept Kuhn's model for the sake of argument. Although I have severely criticized the model on both empirical and normative grounds, it has been sufficiently absorbed into the general, as well as scientific, culture to be taken as a baseline consensus (Fuller 2000b, 2003). But, even given Kuhn's model, two further questions may be posed. Is science becoming more complex? Is it becoming more progressive? These questions are best addressed in steps. Perhaps the least controversial conclusion is that the historical course of science is *asymmetrical*, for however the history is told, we end up in a different place from where we began, especially in terms of fundamental categories. A more controversial conclusion is that this history is *irreversible*. In other words, there is no straightforward way, given where we are now in the history of chemistry, say, to resurrect Priestley's vision as a viable alternative to chemistry's dominant paradigm. I regard this as more controversial because it implies that not only Priestley's language but also his ideas are beyond renewed appropriation.

This brings us to the issue of complexity. From one perspective, the history of science looks like a story of reduced complexity, caused by an intensified division of cognitive labor. Thus, Kuhn's account is often read as justifying increased specialization and putting paid to dreams of a Grand Unified Theory of Everything. Certainly, when compared to the corresponding paradigmatic victors, the alternatives that have been written out of the history of science – whether it be Descartes in physics, Priestley in chemistry, or Lamarck in biology – tended to have had a more ambitious intellectual agenda that, by today's lights, did not respect disciplinary boundaries or even the fact–value distinction. But to our eyes, their science looks simply more complicated, not more complex, than ours: They mixed things we now prefer to keep separate, and consequently never developed models whose explanatory breadth were matched by the fine-grained level of prediction and control that our science currently allows. Of course, a less charitable way of characterizing the order exhibited by the metalanguages of successive paradigms is in terms of the ability to say more and more about less and less.

3. Darwin and the Complexity of Nature

So, then, is Kuhn's narrative one of progress – or not? An interesting perspective on this question is provided by the quarter-century dispute between Richard Dawkins and Stephen Jay Gould over the shape of modern evolutionary theory, ended only by Gould's death in 2002 (Sterelny 2001). Kuhn himself frequently alluded to the non-teleological character of Darwin's theory of evolution as a model for his account. Dawkins and Gould agree on the basic facts of evolution – the long-term differentiation of species through descent with modification – but disagree over how these facts are best to be explained, which in turn colors their judgment as to whether any sense of progress can be attributed to evolution.

In terms introduced in the first chapter, Dawkins is a "functionalist" and Gould a "formalist." Dawkins's fixation on what he calls the "gene's eye-view of the world" is a secularized hold-over of theodicy's concern for the *cui bono* of organic survival:

Were nature an intelligent agent (which Dawkins, of course, denies), on whose behalf would it be acting? His answer is the genes, whose endless propagation best makes sense of the course of evolutionary history as it has actually unfolded. In contrast, Gould was more inclined to concede evolution's fundamental irrationality by focusing on the paths not taken to get to where we are today.

Michael Ruse (1996) has persuasively argued that much can be explained here by differences in the British and American contexts for Darwin's reception. Dawkins has continued a trajectory begun by T. H. Huxley, who followed Comte and other positivists in wanting the natural sciences to replace theology as the cornerstone of the curriculum. Indeed, Dawkins has gone one further and taken down humans along with God, which would have horrified Huxley and the positivists. In contrast, Gould's Harvard chair was named for Louis Agassiz, the Swiss student of Cuvier who in the mid-19th century migrated to the United States to organize Harvard's museum of comparative zoology. Even after the success of Darwin's theory of evolution had eclipsed Agassiz's own creationism, Agassiz's skepticism about smooth transitions in the history of life on Earth – epitomized by his seminal idea of an "Ice Age" – has continued to anchor American scientific intuitions.

The nub of the disagreement between Dawkins and Gould turns on what philosophers call the "units" or "levels" of selection (Brandon and Burian 1984). To simplify the argument, let us take "natural selection" to cover all the conditions in the environment responsible for the differential reproductive pattern in species. On what exactly do these conditions – or "forces," if you will – act?

Dawkins, the optimist of the two, believes that natural selection acts on genetic lines, strands of DNA, the contents of which overlap by 90+ percent for all animal species. From what he calls the "gene's eye-view of the world," there is relatively little difference between species – especially when compared with the physical constitution of all matter. For Dawkins (1976), evolution registers a kind of progress, as fundamentally unsustainable forms of life are eliminated over time. The only big miracle, so to speak, was that a stable configuration of micro-molecules managed to reproduce itself billions of years ago to become the first life form. Thereafter natural selection took over, with new species emerging from relatively minor differences in genetic sequences that propagate in response to changes in the environment.

Dawkins's version of natural selection presupposes that biological science has settled on its basic laws but continues to do experiments and collect data to determine how they are expressed under various conditions and constraints. Progress thus comes from intensified prediction and control, a finer-grained fit between representation and reality. As an account of the history of science, you might expect this kind of upbeat story from a positivist for whom the differences between models, or theories, boil down to their empirical breadth and depth, all of which is reducible to logical complexes of observations. That theories appear rather different in what they claim about the world should be treated as a relatively superficial fact of linguistic ornamentation, much as Dawkins regards the differences in the morphology of organisms that are patently obvious to the naturalist's naked eye.

This is in striking contrast with Gould, who comes to evolution with an intellectual background closer to that of Darwin himself, namely, as a natural historian with minimal knowledge of molecular biology. From the standpoint of the field, rather than the lab, species look wondrously multifarious, not marginally different from each other. For both Gould and Darwin, natural selection operates on populations of organisms in local environments. Which ones are eligible to contribute to the survival of their respective species is determined by whether they live long enough to reproduce. Moreover, Gould inherited Darwin's pessimistic attitude toward evolution's main lesson, namely, that natural selection weeds out most potential blueprints for life. Compared with the Cambrian Age, a half-billion years ago, when all the major blueprints emerged over a short period, relatively few radically different designs for life flourish today, though admittedly the surviving plans have become more elaborated vis-à-vis changing environmental conditions (Gould 1989).

Kuhn's account of the history of science is closer in spirit to Gould's than Dawkins's vision of evolution. Kuhn shares Gould's ambivalence toward the normative lesson of evolution, specifically that the increase in complexity in the evolution of extant species or theories has come at the cost of a decrease in complexity, as represented by the range of extinct species or theories. From the standpoint shared by Dawkins's and Kuhn's positivistic foes, the narrowing of evolutionary horizons may look like a progressive arrow toward global sustainability. But for Gould and Kuhn, the arrow betrays a long-term attrition of fundamental biological and scientific alternatives.

4. The Normative Dimensions of Complexity

The normative significance of complexity in both biological and scientific evolution is so radically underdetermined because all sides adopt a *naturalistic* approach to normativity. In other words, everyone wants to derive their value orientation from the preponderance of the evidence – yet without assuming that the revealed empirical patterns reflect the work of some prior intelligence, design, or plan. The result is reflected in the radically contrasting conclusions drawn by Dawkins and Gould about whether evolution constitutes progress.

That a "norm" in a sense more robust than simply "mean expectation" should somehow (perhaps implicitly?) emerge from the normal course of experience is problematic for three interrelated reasons. First, norms presuppose a clear standard in terms of which particular acts or events are deemed desirable, acceptable – or not. Second, from a strictly normative standpoint, there is no expectation that most such acts or events will appear in a favorable light. That most acts or events might fail to meet the norms is not necessarily a strike against the norms, as long as successor acts or events might do so in the future. Third, the norms should constitute an ideal that can serve as a goal toward which one might strive and with respect to which some communication may transpire about the state of one's progress along the way.

For these three reasons, normative theorists – of both science and society more generally – have been suspicious of arguments that presume the legitimacy of tradition or the probative value of inductive generalization. Only if one additionally assumes a pre-ordained harmony between the structure of our minds and that of reality can enduring self-organizing patterns of experience plausibly carry any normative weight at all. To be sure, just this assumption robustly recurs in the history of Western philosophy: sometimes in theistic garb, from Aristotle through Leibniz; sometimes in atheistic garb, from Hume through Wittgenstein. Its most scientifically respectable version is probably the "anthropic principle," which argues that some profound design features of the physical universe follow from the bare fact that humans are in a position to study it systematically (Barrow and Tipler 1986).

Nevertheless, most normative theorists have tended not to be wishful inductivists. Rather, they have been fallibilists, but more importantly also *corrigibilists*, with regard to experience. In other words, for them, experience reveals that we are not merely error-prone but we are capable of correcting that tendency. That failed chemical revolutionary, Joseph Priestley, originated an image, partly inspired by Newton's calculus, designed to quantify the relevant sense of corrigibility, namely, the successive ("asymptotic") approximation of our theories to the structure of reality (Laudan 1981: ch. 14). This image, subsequently developed in the pragmatist and neo-positivist traditions by Charles Sanders Peirce and Karl Popper, presupposes that somehow reality communicates sufficiently the source of our errors – say, as the outcome of an experiment – that each subsequent theory gets us a bit closer to the truth. Certainly such a logic lay behind Sneed and Stegmüller's attempt to rewrite Kuhnian paradigms as a sequence of positivistic metalanguages.

After Kant, philosophers have used the term *intelligibility* to capture the idea that scientific inquiry constitutes an exercise in genuine learning, not simply random guessing (Dear 2006). Built into this term is the assumption that science's search for a Grand Unified Theory of Everything – a task that Kant, like Priestley, Peirce, and Popper, associated pre-eminently with Newton – would be pointless if reality were not the product of a unified intelligence, basically a mind that differs in degree, but not in kind, from our own. Unlike the idea of preordained harmony mentioned earlier, this assumption implies that we and the intelligent designer have distinct minds, albeit of a very similar structure. Science therefore amounts to an extended dialogue with the designer, in which experiments and other empirical tests serve as the mode of address, which in the fullness of time (hopefully) leads to a "meeting of minds."

The nature of this "meeting" has been subject to multiple interpretations. Most philosophers, typically identified with "convergent scientific realism," treat the claim to mean that we come to comprehend the divine plan. However, there is a bolder reading, one certainly advanced by Priestley and probably even intended by Newton. It involves claiming that we acquire the knowledge necessary for *completing* the plan. In this respect, the appliance of science to reform the world through social technologies becomes a moral imperative. (Noble 1997 is the best, though very critical, history of this broadly "positivistic" sensibility, which marks the seamless transition from Christianity at its most medieval to modernity at its most futuristic.)

That the structure of reality might be usefully seen as the communication of meaning is enshrined in modern information theory, including its analogical extension in genetics (Kay 2000). It is thus not by accident, as we saw in the previous chapter, that the leading advocate of IDT, William Dembski (1998), has recourse to "complex specified information" as his foundational principle. In information theory, a standard procedure for extracting the meaning of a message involves the message's receiver in a two-step process of successive elimination: first, remove whatever is shared by all possible messages (i.e. grammatical rules, common words, etc.), and then remove whatever appears arbitrarily related to the overall pattern detected (i.e. random noise). In short, whatever cannot be ascribed to necessity or chance is *ipso facto* meaningful: it has been transmitted by design and is intended for the message's receiver.

This procedure recalls the pragmatic context of information theory's founder, Claude Shannon, an electrical engineer coping with interference in walkie-talkie radio signals emanating from the battlefields of World War II (Fuller 2005). To be sure, as

IDT's many critics have observed, the procedure is not foolproof (Fitelson, Stephens, and Sober 1999). Nevertheless, its heuristic value has been accepted in settings less fraught with theological controversy. For example, it underwrites the interpretive principles that routinely persuade analytic philosophers of our ability to make sense of people in times and places radically different from our own. Perhaps unsurprisingly, the two main statements of these principles, Quine's "principle of charity" and Davidson's "principle of humanity," were inspired by intelligence work undertaken with the aid of information theory during the Cold War (Fuller 1988: ch. 6; cf. Reisch 2005). The most extreme version of this application appears in NASA's Search for Extraterrestrial Intelligence, or SETI, project (Basalla 2006: ch. 9).

A more telling example comes from the neo-Darwinian heartland itself. Evolutionary psychology has been a recent beneficiary of "informatization," as it negotiates a middle way between the social sciences' own version of the necessity/chance binary: on the one hand, biological determinism (whereby humans would respond similarly, regardless of differences in the environment) and, on the other, sociological relativism (whereby human response would simply mirror differences in the environment). Thus, corresponding to Dembski's idea of complex specified information, there has been a revival of a crypto-normative sense of "human nature," now in terms of our capacity to respond to environmental changes in various ways but always within a genetically circumscribed behavioral bandwidth (e.g. Pinker 2002). Darwinian terms-of-art for these limits, such as "adaptiveness" and "fitness," imply, in Dawkins's inimitable phrase, "design without a designer" – whatever that means.

Ironically, Dawkins and other "Darwinian fundamentalists" like Steven Pinker (2002) and Daniel Dennett (1995) seem to end up unwittingly investing as much agency, perhaps even intentionality, in "natural selection" (thereby resurrecting the expression's metaphorical roots in animal husbandry) as their creationist foes. Not surprisingly, this tendency led Gould to coin the term "exaptation" to capture the opposing idea that whatever natural fit appears to obtain between an organism's behavior and its environment is really just a by-product of some other trait that facilitated the survival of previous generations of the organism. In this way, Gould tried to preserve the radical contingency of evolution and the precarious grip that organisms ultimately have on their selection environment.

Although Darwinists rounded on Gould for underselling the explanatory power of natural selection, and more generally for making evolution seem like an irrational process, he was much closer in spirit to Darwin than are today's so-called Darwinian fundamentalists. At least, Gould shared Darwin's original clarity about what is entailed by rejecting the idea of design in nature: It involves something more substantial than simply giving design a new name and obscuring its source. When one of Darwin's youthful inspirations, David Hume, rejected design as a premise for justifying the existence of God, his skepticism cut deeper than simply severing the epistemic authority of the natural sciences from that of theology. His argument was also meant to deflate Newton's own theistically inspired pretensions of having penetrated nature's causal mysteries. At most, Hume believed, Newton had grasped empirical regularities of great import to the amelioration of the human condition. But Hume refused to impute more than that to Newton's achievement. Something similar might be said of Darwin's own modest sense of intellectual achievement.

Significantly, Darwin and even Gould felt much more comfortable with the job title of "natural historian" than "natural scientist." For them, the array of species that characterize what is still figuratively called "the tree of life" revealed natural selection to be generative of complication rather than complexity, strictly speaking. Were Darwin teleported to one of today's cutting-edge biotechnology labs, he would be astonished to learn of the medical benefits to be had from "xenotransplantation," which involves the functional transfer of an organ or gene between members of different species. The efficacy of this procedure supports the view that organisms are designed like machines from a few overlapping blueprints that permit some interchangeable parts (Smith 1998). The likelihood that a chance-based self-organizing process would result in a pattern of species whose biochemical basis is sufficiently similar and stable to allow for so much successful xenotransplantation is probably such that Darwin – and perhaps even Gould – would have thought it unlikely ever to be realized. Yet not only have cross-species substitutions been realized, but they are likely to increase in the future.

Darwinian fundamentalists fail to understand that Darwin divested design of its divine origins mainly because he could not find enough good done over the course of natural history to justify the existence of a benevolent God. Darwin held God to anthropic ethical standards and found the actions that natural theologians imputed to the deity wanting, once the evidence was considered in its totality. But Darwin was writing in the mid-19th century, not the early 21st century. This suggests the following counterfactual prospect: just as Darwin became dissuaded of design by considering the pointlessness of the mass extinctions registered in the fossil record, which only came to be accepted in his life-time, his faith in design might be rekindled, were he to see just how much control the biomedical sciences have enabled us to exert over life processes. Complication may turn out to reveal genuine complexity, after all. We shall revisit this point at the end of this book.

5. Conclusion: Computerized Evolution as Intelligently Designed Complexity

Counterfactual speculations about Darwin's attitudes toward today's feats of biotechnology may appear fanciful, but an even clearer revival of design-based cosmological thinking may be found in the recent enthusiasm for the computer modeling of evolutionary processes. Charles Babbage, the computer's inventor and a successor to Newton's Cambridge chair, first raised the prospect in 1838 – in an explicit attempt to update design-based arguments for God's existence (Knight 2004: 49–51). That the deity might have created the universe with stochastic features to enable humans to exercise a measure of free will was widely seen . as belonging to Priestley's radical vision of God empowering us to extend and perhaps even complete creation. What appeared striking to the 19th-century mind, but had come to be forgotten by the mid-20th century, is that chance-based processes could be embedded in a cosmic design that a computer might be programed to simulate. The lost implication that Babbage wished to stress was that God might be the cosmic programer who deliberately inserted elements of chance.

Von Neumann may be seen as having deepened Babbage's original point with the benefit of nearly a century of Darwinian evolutionary thought. In 1949 he proposed how it might be possible to build self-reproducing automata programed to generate heritable mutations, the outputs of which mimicked the complexity of natural history. In effect, these machines would be capable of randomly changing their basic structure, but in ways that enabled them to reproduce without degenerating in size or level of organization. As Mirowski (2002) has observed, von Neumann's idea quickly caught hold in economics as the framework for regulating markets – both locally and globally – in a volatile postwar world. Von Neumann was read as having successfully answered the skepticism of Austrian economists like Friedrich Hayek and other laissez-faire enthusiasts, who doubted that the complications resulting from the natural evolution of markets could be managed as what might be called "Keynesian complexity."

The spirit of von Neumann lives on at the Santa Fe Institute, especially its resident guru, the biophysicist Stuart Kauffman (1995), who coined the expression "order for free" to capture the complexity exhibited by bootstrapped Darwinian life forms whose fitness always transpires "at the edge of chaos." The computer simulations produced by Kauffman's colleagues and fellow-travelers, across a wide range of disciplines from physics to economics, have approximated the parameters of known evolutionary processes to an impressive degree, though without yet having specified the physical conditions of life's origins. Central to the concerns of the Santa Fe Institute has been an issue that faced Darwin's theory of evolution at its outset, especially given its stress on natural selection acting on chance variations: suppose we rewound Earth's history and replayed it on a computer simulation. Would radically different life forms have emerged over a comparable time frame, or would they have to conform to the general blueprints known to have emerged over the course of natural history?

Among Darwin's contemporaries, the renegade Catholic biologist St George Mivart argued for the latter position, going so far as to suggest that natural history provides evidence for convergence over time in the morphology of life forms. *Contra* Darwin, Mivart interpreted mass extinctions optimistically as implying that evolution is sufficiently "intelligent" to display self-correcting tendencies (Brooke 1991: 283–4). Although eventually ostracized by his teacher T. H. Huxley and excommunicated by the Church, Mivart neatly reconciled divine will and natural law: God need not have created life at all, but once he decided to do it, it had to be done a certain way (or within a certain range of ways). By analogy, I may or may not want to play a game, but once I decide to play, there is a right and wrong way to do it. With respect to the "meaning of life," it follows that the religiously inspired scientist stands in the same position as an anthropologist trying to fathom why I decided to play the game, which involves understanding how the game realizes my intention in setting it up.

Mivart's strategic amendment of Darwin is clear even without his own theological gloss. Any computer simulation of life processes needs to be front-loaded with appropriate constraints to increase the chances that a random event will evolve into viable life forms over a long period and across the entire Earth. In this context, one starts to attach the term "theory" to "complexity," which, along with "chaos theory," is supposed to provide the mathematical foundations for what former *Scientific American* editor John Horgan (1996) has called the science of "chaoplexity." For Horgan, the significance of these Santa Fe-driven developments is symptomatic of the increased "virtualization" of scientific inquiry, first from the field observation to the laboratory experiment, and now from the laboratory experiment to the computer simulation. While agreeing with Horgan's general analysis of the situation, it is nevertheless useful to distinguish what *chaos* and *complexity* each bring to both the scientific and religious dimensions of explaining life's origins.

Chaos refers to the mathematical properties of phenomena whereby an initial seemingly random event triggers a chain of consequences that result in a permanent shift in the equilibrium of a system. One obvious application of chaos theory is Gould's "punctuated equilibrium" account of evolution, whereby some improbable disturbance so radically changes the selection environment that over a short period an entire set of species is replaced by another (i.e. the differences in survival rates between the sets is reversed). Chaos can also model other versions of "catastrophism," including accounts of creation that make room for, say, floods, miracles, and other event-like divine interventions. Here it is worth stressing that what traditionally distinguishes a miracle from an ordinary random event in the Abrahamic religions is the miracle's capacity to change the *normal* way people behave toward each other, the world, and so on. To be sure, this radical change may be largely a self-fulfilling prophecy, as people behave differently as a result of believing that an event is a miracle. Nevertheless, whether the event is explained supernaturally as a miracle or naturalistically as a self-fulfilling prophecy, it can be modeled by chaos theory.

The work of Kauffman and the Santa Fe Institute is better captured by complexity theory, the conceptual complement of chaos theory. If chaos theory models how an apparently stable set of phenomena can yield instability and ultimately radical change under specific "miraculous" conditions, complexity theory models how an apparently random set of phenomena may reflect a deeper "design," a sense of order whose recurring properties can be captured by computer simulations, including ones that execute "self-interacting" programs, i.e. capable of changing themselves over time in response to their outputs. In other words, the conception of "design" implied in complexity theory is quite dynamic, as Babbage envisaged on theological grounds, but is certainly not an *a priori* blueprint, as design was conceptualized by, say, Newton and Paley.

Unsurprisingly, perhaps, economists interested in modeling capitalism's elusive "invisible hand" have been drawn to complexity theory. But also biologists unsatisfied with Darwinian fundamentalism have been trying to merge chaos and complexity theories in an "evolutionary developmental" synthesis (e.g. Carroll 2005). "Evo-devo" theory basically gives a grounding in modern genetics and molecular biology to Gould's view of natural history.

Its guiding intuition is that the evolution of species on Earth, while continuing to display enormous variety, has nevertheless remained within a narrower range than would be expected by appealing simply to changing features in the physical environment. Thus, evolution is constrained by something more systematic than simply the vicissitudes of natural selection.

Intelligent design theorists have unwittingly revisited Babbage's original motivation for the construction of Santa Fe-style computer simulations. While director of the Michael Polanyi Center for Complexity, Information and Design at Baylor University, Texas, Dembski (2001) speculated that a future simulation of life's origins would display sufficient "specified complexity" to provide evidence for a divine creator. But by Babbage's lights, Dembski has set the standards of proof rather high, since the very prospect of simulating evolution on a computer to the satisfaction of someone like Kauffman already bolsters the case for a divine creator. After all, any such computer program is, strictly speaking, a product of intelligent design, not literally a self-organizing entity surviving at the edge of chaos. If humans can program a computer that generates an output with such deep self-organizing properties, why couldn't God? In short, the status of intelligent design as an alternative explanation for the emergence of life is likely to become more heated as evolutionists rely increasingly on computers to demonstrate that natural history is not merely complicated but genuinely complex. This is simply because the two positions will become harder to distinguish from each other, and the evolutionists will be playing on the intelligent design theorist's turf. The alternative, of course, would be for evolutionists to demonstrate the existence of a von Neumann machine in the wild that bears no signs of design, human or otherwise.

4

America as a Legal Battleground

1. Expertise on the Nature of Science

In February 2005, the Thomas More Law Center of Ann Arbor, Michigan, asked me to serve as a "rebuttal witness" for the defense in *Kitzmiller v. Dover Area School District*, a trial scheduled to begin early in the autumn of that year, which would be the first case to test the eligibility of "intelligent design theory" (IDT) for inclusion alongside the neo-Darwinian theory of evolution in high school biology classes. As a rebuttal witness, my charge was to contradict the claims made by the plaintiffs' witnesses, all of whom were seasoned veterans of related trials involving creationism. Their opinions were already compiled (Dembski and Ruse

2004) and their biographies featured on the website of the plaintiffs' legal team, the American Civil Liberties Union (ACLU). I decided to participate without knowing this background but simply after having read the expert witness reports as filed by the plaintiffs' lawyers. These struck me as based on tendentious understandings of the nature of science that would not have survived scrutiny on an informed listserv like HOPOS-L (dedicated to the history of the philosophy of science), let alone the peer review process of a relevant journal. I concluded that the plaintiffs' experts simply took advantage of their "expert" status to offer their sincerely held but professionally uncensored opinions.

My critical eye was clearly informed by knowledge gained from science and technology studies, since, while having written on IDT's struggle for scientific legitimacy (Fuller 1998b), I am not an advocate of – or expert in – either IDT or, for that matter, neo-Darwinism. But my *prima facie* marginality to the positions under dispute did not deter me. Indeed, I may be the first person to declare under oath that someone trained in the history, philosophy, and sociology of science – that is, science and technology studies (which I glossed as "social epistemology" during the trial) – can better evaluate the scientific standing of a field of inquiry than someone formally trained in science.

Of course, scientists trained in particular fields are better placed to issue judgments about the epistemic standing of claims and practices in those fields. But that is different from the second-order problem of whether such fields constitute *sciences*. A great virtue of the US Constitution is its delegation of educational decisions to the local level, so that those who pay for the schools (through taxes) get to govern them (Fuller 2006a: 174–9). There is no educational ministry that sets a "national curriculum," as commonly found in much of the developed world. As long as enough students pass state-based (i.e. based in a particular US state) subject exams, the curriculum can be organized as the locals see fit – that is, until a legal challenge is raised.

Letting the locals decide science education simultaneously embraces two points about science to which the field of science and technology studies has been sensitive (cf. Fuller and Collier

2004). First, "science" mainly refers not to narrowly circum-scribed technical competence, but to knowledge whose aspiration to transparent accountability and universal applicability can autho-rize thought and action in the public sphere. If one considers the wide array of practices that pass as, say, "social science" or even "biological science," it becomes obvious that whatever qualifies these fields as sciences will *not* be found in some common core activity, but rather in how multifarious activities are interrelated and regarded by the larger society (Fuller 1997: ch. 2). Second, relatively few of those taught science at any level become practic-ing scientists, yet everyone takes decisions that involve the evalu-ation of scientific claims. I may not be able to tell a lab geneticist how to determine the mutation rate of fruit flies, but then what makes what the geneticist does scientific is not actually deter-mined in her lab, but by how it is received – or "generalized" – in the wider scientific community and society at large.

The role of the reception environment for practices achieving scientific status should not be underestimated. It would be easy to imagine a world in which scientists do the same work but it is treated as simply a game. It would still require all the same training and skill, but it would not be the basis for technology, policy, or any other socially significant innovations – or, for that matter, metaphysical conclusions. After all, Galileo refashioned the telescope from a child's toy, and a while had to pass before people took that as having much significance. In that respect, to insist on a purist definition of science – one ultimately determined by and for practicing scientists – is to reduce its significance. Much of probability theory was originally developed in the mid-17th century to help idle nobles win card games and budding capitalists to make prudent investments. As it happened, it was soon taken up to construct "vital statistics" in the form of birth and mortality rates, which enabled states to project themselves as administering to a social organism (Hacking 1975). But another 200 years had to pass before James Clerk Maxwell declared, in the 1873 meeting of the British Association for the Advancement of Science, that physics is irreducibly statistical (Porter 1986).

To be sure, I was not the first expert witness in a science-based trial whose expertise lay beyond the sciences formally under contention. The testimony of Michael Ruse, a scientific amateur, was the intellectual centerpiece in the verdict delivered in *McLean v. Arkansas Board of Education* (1982). Although the topic of Ruse's Ph.D. thesis (University of Bristol, 1970) was "The Nature of Biology," all of his degrees have been in philosophy. His pedigree typifies the first generation of philosophers of science whose intellectual center of gravity is biology (others include David Hull, Alexander Rosenberg, Elliott Sober, Philip Kitcher, and lately Daniel Dennett). The previous two generations of philosophers of science (spanning the transition from the logical positivists to Kuhn and the historicists) had received formal training in physics and then migrated to philosophy to keep alive the broad cognitive interest in natural philosophy that scientific specialization had abandoned. The shift in the direction of migration ("into" vs "out of" philosophy) reflects a repositioning of the philosopher of science from legislator to underlaborer, a point originally developed in Fuller (2000b: ch. 6).

From this standpoint, Ruse is a transitional figure who proffered a physics-based definition of science and whose writing generally brings the whole weight of the history of science – not merely current scientific judgment – to bear on debates over Darwinism (Ruse 1983). Notwithstanding his pro-Darwin credentials, Ruse has never been a devotee of the "newer is better" sense of intellectual progress sometimes found in contemporary philosophy of science touched by biology, which amounts to a Darwinian fundamentalism. Ruse's intellectual independence was recently highlighted in a nasty email exchange with Dennett, who holds just such a view (Brown 2006).

Ruse had made his name with a book entitled, *The Darwinian Revolution: Science Red in Tooth and Claw* (Ruse 1979), a philosophical history that aimed to extend Kuhn's influential theory of scientific revolutions from physics to biology. Ruse testified to the presence of "demarcation criteria" by which one can distinguish science from non-science. I shall discuss this approach more fully below, but suffice it say that Ruse argued that the proposed version of creation science failed to be a science, because it relied on unfalsifiable – indeed, infallible – biblical pronouncements about nature. Ruse's testimony was gone over with a fine-tooth comb by his philosophical colleagues, most of whom, while stopping short of accusing him of perjury, complained that he had advanced a largely discredited conception of science (e.g. Laudan (1982); for my own ambivalent support for Ruse, see Fuller and Collier (2004: 198)).

Nevertheless, all things considered, the precedent set by Ruse's testimony in *McLean* was the right one, and it is unfortunate that the judge in *Kitzmiller* made no reference to it, though he referred to the case in which it figured. Instead, his reasoning drew on the US Supreme Court decision in *Edwards v. Aguillard* (1987), which defined the distinction between science and religion in terms of *motive* rather than *method*. The difference matters, even though – so far – creationism and IDT have largely failed to meet either the *McLean* or the *Edwards* standard. If we judged scientific theories mainly by the motives of their advocates, then certainly 100 or even 50 years ago (some might say, even now), we would have disallowed Darwinism because its promoters have been motivated by classism, racism, and sexism. A great strength of the scientific world-view is precisely driving a wedge between motive and method, so that the validity of ideas need not depend on the groups that originate or more obviously stand to benefit by them. Mainstreaming IDT would have exactly that consequence, as it did in the case of Darwinism.

In contrast, a motive-based definition reinforces a false dichotomy between *science* and *religion*, while obscuring a genuine distinction in the contexts of *discovery* and *justification* in science. The former refers to the various theoretical motivations – metaphysical, religious, political – for doing science, the latter to the validation of the actual knowledge claims (Fuller 2003: ch. 16). Two contexts work best when they work apart, a "separation of powers," to recall the language of the US Constitution. While the validity of a scientific theory ultimately rests on its acceptance by those who do not share the motives and commitments of its originators, it is equally true that only those with certain motives and commitments would have found the theory plausible to pursue in the first place. This latter point, concerning the context of discovery, is crucial for *Kitzmiller* and related cases, which are more directly about the *teaching* than the *testing* of science knowledge claims.

Most legal definitions of science appeal to "testability," and it is easy to see why – not because testability underwrites truth but because it raises the confidence of those absent from the original research site who may wish to build on what was done there. The burden of proof is shouldered by those who claim that a demonstration should be accepted by those not present to witness it. In that respect, scientists enjoy a fiduciary relationship with the public that goes beyond their guild norms of what counts as good scientific practice. Principal-agent theory in economics arguably captures the spirit of that relationship (Turner 2003: ch. 4). The question, then, for someone with my alleged expertise is whether particular scientists or groups of scientists – such as Darwinists or ID theorists – abide by the terms of the fiduciary relationship in specific settings: in this case, whether the public interest in science is served by excluding IDT from the science curriculum.

Respecting the two contexts, IDT's two leading proponents, Michael Behe and William Dembski, have repeatedly stressed both the theological inspiration of their work and its ultimate validation by scientific means. For the most part, their academic critics have also dealt with their work in the context of justification without bringing in its religious context of discovery. This is in sharp contrast to the state of play at the time of *McLean*. Here one need only contrast Nelkin (1982) and Ruse (2003), important books published shortly after *McLean* and before *Kitzmiller*. Both are clearly pro-evolution and anti-IDT but the nature of that opposition has clearly evolved. Whereas Nelkin situated IDT amidst the scientific illiteracy of biblical literalists, Ruse did so against a broader tradition of natural theology that is admitted to intersect at many points with the pursuit of mainstream physical and biological science. Moreover, Ruse spends much time distinguishing IDT positions from cognate non-IDT positions, arguing against the former on much more technical grounds than was needed 20 years earlier – often by appealing to recent scientific developments (e.g. computer-based models of self-organizing systems) that are still controversial in their own fields.

This striking transformation in the presentation of IDT by its opponents suggests that even evolutionists concede that IDT has become more scientific in the pursuit of its inquiries, thus requiring critiques that approximate the level of sophistication that evolutionists normally would reserve for their more orthodox scientific colleagues. The critiques do not concern IDT's motivating metaphysical assumptions about a divine creator, but rather the validity of specific inferences that IDT proponents draw from bodies of evidence or lines of reasoning already familiar from a variety of independent research contexts (e.g. Fitelson, Stephens, and Sober 1999).

The *Kitzmiller* ruling took a step back from the historical impetus for proposing a "scientific method," from Francis Bacon's *experimentum crucis*, through the versions of the demarcation criteria advanced by the positivists and the Popperians, up to recent sociological preoccupations with boundary construction and maintenance in knowledge production. All of these projects have shared an interest in finding a neutral ground for the adjudication of contesting knowledge claims. To be sure, the projects define "neutrality" rather differently (sometimes the exemplar of neutrality is the judge; sometimes it is the sociologist of science), but they agree that it means something other than simply deferring to those currently called scientists to decide who else can count as scientists. A major exception to this consensus of opinion about the scientific method is provided by the work of Thomas Kuhn and others in science and technology studies who share the late Wittgenstein's anthropological sensibility, whereby science is simply what scientists do. The judge's ruling for the plaintiffs in *Kitzmiller* can be easily understood in these terms.

To put this matter in perspective, it is worth recalling that Francis Bacon was a philosophically minded lawyer who sought a standard for weighing the merits of competing scientific claims in the early 17th century, a period fraught with religious tensions that eventuated in the English Civil War. Bacon's legacy was the "crucial experiment," a watered-down version of which persists in legal definitions of science as "testability." An important feature of the testability standard stressed by Bacon and later philosophers is that it should be possible to evaluate the scientific merits of competing knowledge claims without presupposing the theoretical commitments of any of the claimants. Indeed, this extension of the idea of procedural justice has come to be popularly seen as synonymous with science: the need to present evidence and reasoning fairly without assuming the outcome.

Based on Ruse's testimony in *McLean*, the judge did not simply defer to the experts in the contested field to authorize his decision (Overton 1983). He sought an independent standard, which Ruse provided, in that the falsifiability principle does not beg the question in favor of a science's dominant paradigm. Of course, it helped that Ruse's independent standard also coincided with expert scientific judgment, but the legal precedent had been set. In principle, at least, future candidates for inclusion in the science curriculum could satisfy criteria that do not require the approval of those who claimed to speak on behalf of the entire scientific community. Ruse's testimony seemed to carve out a distinctive role for the "metascientfic" expert on science, in spite of professional philosophical misgivings.

In sum, Judge Overton's endorsement of Ruse's testimony provided a standard of fairness, which I have elsewhere called *epistemic justice* (Fuller 2007: ch. 5). Such a standard operates in the spirit of, say, job qualifications that a candidate meets, despite the personal judgment of a prospective employer, against whom a charge of discrimination could be legitimately raised. The relevance of this point is that a widely reported feature of my testimony was my claim that, given the hegemony that neo-Darwinism enjoys in contemporary biology, an "affirmative action" policy would be necessary for IDT to develop its research program sufficiently to mount a credible challenge. Such a policy may include giving "equal time" to the lesser theory at the high school level. Tales of so-called scientific underdogs who eventually triumph typically either took place before the era of Big Science or do not propose as radical a reorientation of domain of scientific inquiry as IDT does. In any case, it should be always an open question for high school educators, even in science, whether the next generation is exposed exclusively to the received wisdom of the current generation or whether the seeds of epistemic change are planted as well.

To be sure, my testimony included criticism of IDT, as I used the expert witness role as an opportunity to provide a model for how the judge should weigh the merits of both sides of the case. For my trouble, I was quoted authoritatively in the closing arguments by the lawyers on *both* sides. The judge cited me a dozen times in his ruling, unsurprisingly, from whenever I had said something revealing or critical of IDT. In the judge's mind, the case turned on whether IDT is "really" religion *and not science.* Indeed, this forced choice of "either religion or science" was the one constraint to which my testimony had to conform, in the eyes of the defense lawyers. However, my own view – which I have argued throughout this book and believe accords well with those who have studied the matter seriously – is that science and religion are *not* "separate but equal," as the *Kitzmiller* verdict suggests, but rather are substantially overlapping modes of inquiry. The spirit of that position remained present in my testimony, which gave Judge Jones the opportunity to use it to bolster an *exclusively* religious interpretation of IDT.

2. Supernaturalism as Figment and Fact in the Naturalistic Imagination

Let me first declare that I am long on record as a naturalist, though a "reflexive" one who would have naturalists naturalize their own beliefs – that is, to understand the socio-historical conditions that have enabled their own doctrine (Fuller 1992, 1993). As I observed at the start of this book, these conditions were already understood by "Darwin's bulldog," Thomas Henry Huxley, whose Romanes Lecture of 1893, "Evolution and Ethics," argued that, as a matter of historical fact, the triumph of modern science is indebted to the monotheistic religions, which elevate humans to "the image and likeness of God," capable of grasping in a detailed and comprehensive manner the unity of nature for purposes of transforming it according to human needs and purposes. In contrast, the more naturalistic world-views that emerged from the Greco-Roman pagans and the great Eastern religions tended to promote a fatalism that discouraged the industry required for science. Huxley concluded by presenting his largely bewildered secular late-Victorian audience with a paradox: from a naturalistic standpoint, naturalism itself promotes science only after monotheism has inoculated inquirers against naturalism's own self-deflating tendencies. I believe that here Huxley was exactly right.

Nevertheless, the plaintiffs' experts in *Kitzmiller* uniformly claimed that science has progressed by rejecting supernaturalism in favor of naturalism. "Supernaturalism" evokes strong feelings in modern secular culture, as it trades on associations with mysterious spirit-worlds that appear to defy the laws of physics, otherwise known as "animism." This is the form of supernaturalism originally associated with so-called primitive cultures that "failed" to draw a distinction between profane and sacred space − that is to say, between the natural and the divine. To the 19th-century anthropologist, these cultures appeared to inhabit fantasy worlds in which figments of the imagination freely interacted with physical entities. This Victorian attitude has not completely disappeared from so-called cognitive anthropology, which borrows heavily from recent evolutionary psychology (Boyer 2001).

But note that in this context "supernaturalism" is a misnomer. It is the anthropologist − not the putative primitive − who has a vested interest in cordoning off the "supernatural" as a specific realm of being. The primitives' cognitive "problem," then, is that they cannot tell where the anthropologist's sense of "natural" ends and where that of "supernatural" begins. This "problem" continues to haunt discussions of IDT, as neo-Darwinists insist on drawing distinctions where IDT proponents see only continuities.

Two matters are frequently confused in charges of IDT's "supernaturalism." On the one hand, it is true that IDT wishes to pursue research that might eventuate in design-based explanations of the natural world that do not conform to the naturalistic metaphysical assumptions of contemporary biological science. On the other hand, IDT proponents do not regard such explanations as requiring a sharp break with the methods of science. IDT proponents argue primarily by appeal to empirical evidence gathered in the laboratory and the field, employing methods of reasoning − both qualitative and quantitative − familiar from the other branches of science. The only difference here from neo-Darwinists is that IDT proponents tend to be much more conservative about the transferability of conclusions from the lab to the field, or – notably in controversies surrounding the fossil record – from the present to the past. But this "conservatism" is familiar from the annals of positivist, and more broadly instrumentalist, philosophy of science. For example, one of the great modern theorists of scientific experimentation, the Catholic physicist Pierre Duhem (1954), regarded the laboratory as a place that literally manufactures phenomena to theoretical specification whose relevance to pre-existing natural processes remains open. Thus, he saw a clear disciplinary boundary between physics and biology, the former being explicitly artificial and the latter explicitly natural. Moreover, Duhem's conservative attitude is familiar in the social sciences, where, say, success at simulating market behavior on computers or in a laboratory is not necessarily taken as indicative of the causal mechanisms underwriting economic regimes in history.

And, while some IDT proponents also wish to draw corroborating testimony from the Bible, the Bible appears as a heuristic to focus more conventional forms of scientific inquiry (e.g. to pay more attention to radical breaks in the fossil record), not to overrule or undermine such inquiry. Indeed, two features of Newton's world-view that are easily taken for granted today would be difficult to motivate had Newton not imagined himself as trying to simulate the divine thought processes in the spirit promoted in our own time by, say, Hawking (1988).

The first feature is the very idea that there is a unitary view of the world that can be described in terms of a finite set of universal laws. For Newton, these laws were applied in absolute space and time, which he characterized as God's "sensorium," or, as we might now say, the "interface" between God and creation. In other words, Newton imagined that there was a fixed point from which everything could be seen and its overall structure discerned. This is not a trivial assumption. It is much more specific and focused than vague naturalistic appeals to "curiosity" that would portray Newton as a sophisticated version of an animal scoping out its immediate environment.

The second relevant feature of the Newtonian world-view is the search for ultimate explanations that both transcend and unify the disparate phenomena revealed to our immediate senses. Newton sought to provide the same explanation for things that had been previously explained separately. Not surprisingly, the proposed explanations turn out to have a lower prior probability – i.e. they are counterintuitive. But why would Newton have supposed the need for such explanations in the first place? Indeed, it was this search for ultimate explanations that led him to postulate a universal force of physical attraction called "gravity" which, despite its mathematical specifiability, was treated as "action at a distance" and hence "supernatural" by most of his contemporaries. Clearly Newton imagined that the universe was created from a single plan – as did Darwin when he began the intellectual trajectory that led him to the theory of natural selection.

My point in recounting this history is not to demonstrate the scientific validity of religious, let alone all supernatural, beliefs. Rather, it is to highlight the heuristic function of certain religious beliefs – specifically monotheistic ones that privilege humanity – in putting one in a frame of mind that motivates the sustained pursuit of scientific inquiry. That IDT might discourage students from thinking scientifically about the world is a major pedagogical source of concern in *Kitzmiller* and related cases. This suspicion is historically unfounded. On the contrary, design-based thinking fosters the context of scientific discovery. As Darwin's own biography demonstrates, the eventual scope and power of his theory of natural selection was very much the product of a thorough engagement with the argument from design in natural theology. Without this engagement, it is unlikely that Darwin would have arrived at a theory as detailed and nuanced as he did, which has enabled it to serve as a blueprint for subsequent biological research.

No heuristic has consistently stimulated scientific thought as much as the idea of the universe as a machine that can be taken apart and put back together, perhaps even with improvements. The Scientific Revolution arose in the midst of the Protestant Reformation, which had encouraged people to take literally the biblical idea that they were created "in the image and likeness of

God." This prompted an unprecedented number of fruitful scientific labors, culminating in Newton's work, to reverse engineer creation by discovering the creator's underlying "mechanisms." As we have seen in the previous chapters, much of the subsequent history of modern science can be told as the extension of this design-oriented mentality, typically by nonconformist monotheists who would dare "get into the mind of God" or even "play God." They defied the church without denying their religion. This side of the history of biology is best told not through Darwin's own field of natural history but through genetics, which finally provided a mechanistic underpinning to the originally vague process of natural selection. Here the two leading figures were devout Christians, Gregor Mendel and Theodosius Dobzhansky. Moreover, the design mentality persists today in trends toward treating physical reality as the output of a computer program – an idea that the computer's inventor, Charles Babbage, originally floated as an argument for God's existence in the 1830s.

When criticizing IDT, neo-Darwinists are inclined to slide from observing (correctly) that IDT challenges the metaphysical naturalism of contemporary biology to inferring (incorrectly) that IDT challenges the established methods of scientific inquiry. It is the latter, not the former, point that conjures up the image of "supernaturalism" as equivalent to a primitive "irrationalism." However, IDT does not challenge science, only the artificially restricted conceptual horizons within which science is practiced under the neo-Darwinist regime.

The history of science is full of hypotheses of "supernatural" inspiration, in that the hypothesized entities are not observable in the normal run of experience, but only under specially crafted conditions that are unavailable when the hypotheses are first proposed. In short, the right clever experiment has yet to be designed. Typically, these supernatural hypotheses – expressions of what is less prejudicially called "metaphysical realism" – receive their initial grounding in a mathematically significant pattern that points to a deeper level of explanation. In the case of, say, Newton's appeal to gravitational attraction or Mendel's to hereditary factors, these hypotheses have had theistic origins that survive in contemporary IDT – namely, ideas of a divine plan and special creation. Science, on this view, is part of the project of creating "a heaven on Earth," to recall the old Enlightenment ideal, specifically by realizing in matter things that first exist as products of the (here mathematical) imagination, which is itself presumed – as Newton did of his own mind – divinely inspired (the *locus classicus* of this interpretation is Becker 1932).

However, the role of the supernatural in the advancement of science tends to be erased once experimental proof is found for these originally mysterious entities. The context of justification is effectively read back into the context of discovery, which then makes naturalism look persuasive. Newton's and Mendel's theistic inspirations are relegated to superfluity, since subsequent experiments and attendant reasoning are sufficient to demonstrate gravity and genes. Naturalism is thus the philosophical concomitant of a Whig historiography of science, whereby the past is seen in retrospect (from where it ended up), not in prospect (from where it emerged). The interesting exception, as one might expect, is neo-Darwinism itself, where "natural selection" has figured in less than 1 percent of the abstracts of all biology articles indexed since 1960. Even "evolution," in all its protean uses, has figured in only 12 percent of them. Does this mean that the reported research is sufficiently self-contained that it can be explained without appeal to such higher-order concepts? Apparently not. Instead we are told that neo-Darwinism has become so fundamental to the conduct of the life sciences that it is always already assumed. We shall return to this issue in the next chapter.

3. Philosophical Resistance to Scientific Naturalism

IDT's attempt to embrace a philosophy of science that extends beyond naturalism does not reflect the eccentricity of a reactionary scientific movement. On the contrary, it probably represents the mainstream opinion of philosophers themselves. Despite the many philosophers of science who over the years have volunteered their services to the neo-Darwinist cause, naturalism remains a controversial position within academic philosophy. It remains a minority position in philosophy as a whole, though it may enjoy majority status in the philosophy of science – understood as a field distinct from and more specialized than epistemology, the general theory of knowledge. Here it is worth recalling why philosophers have been traditionally disinclined from submitting to the authority of science on matters of knowledge. A good entry point is the shifting status of *naturalism* as a doctrine in the history of philosophy.

Like most broadly defined philosophical positions, naturalism has been interpreted in various ways ever since the term was coined 300 years ago to characterize the philosophy of the pantheistic rationalist, Baruch Spinoza (Israel 2001). Specifically, naturalism has metamorphosed from a heretical to a dogmatic attitude toward knowledge (Fuller 2004). It originally enabled science to challenge theological orthodoxy. But today naturalism is the voice of the scientific establishment, especially when dealing with theologically inspired interlopers, such as IDT. A candid expression of this viewpoint is provided by Richard Lewontin, the Harvard population geneticist who has spent much of his career criticizing illicit extensions of neo-Darwinism into matters of public policy:

> It is not that the methods and institutions of science somehow compel us to accept a material explanation of the phenomenal world, but, on the contrary, that we are forced by our *a priori* adherence to material causes to create an apparatus of investigation and a set of concepts that produce material explanations, no matter how counter-intuitive, no matter how mystifying to the uninitiated. Moreover, that materialism is absolute, for we cannot allow a Divine Foot in the door. (Lewontin 1997: 31)

What follows is a necessarily brief history of this transformation in the fortunes of naturalism.

Spinoza controversially held that the natural and the divine present two aspects of the same reality. We literally inhabit God's being. The deity does not reside in a superior reality beyond the one with which we are normally acquainted. In this form, naturalism encouraged theologians to advance historical and empirical explanations of biblical phenomena, with God revealing himself through the usual methods of science. Indeed, Spinoza's naturalism largely inspired the separation of academic theology from its traditional pastoral function, a seminal moment in the modern understanding of "knowledge pursued for its own sake," and what Max Weber later called "science as a vocation."

At first, naturalism was politically controversial because it denied the legitimacy of claims to divinely sanctioned authority that could not be backed with secular evidence, such as properly validated documents. However, naturalism became controversial within the academy once the methods of science began to sharply deviate from the documentary base of theology – which coincides with when the natural sciences started to challenge the authority of the humanities in the universities in the mid-19th century. More generally, this corresponds to the ascendancy of laboratory- and field-based over text-based research. By the mid-20th century, "naturalism," now understood in its contemporary sense as genuflection to the natural sciences, became the dominant academic ideology, in the sense that the natural sciences – more specifically the physical sciences – established the benchmark against which other forms of knowledge came to be judged. Moreover, this has extended beyond core epistemic judgments of empirical reliability to size of research grants, proliferation of publications, and intensification of technicality.

However, most philosophers have resisted the charms of naturalism, mainly because in practice it would hand over epistemic authority to a specific community of inquirers – scientists and their authorized agents – who are no less prone to errors of judgment than non-scientists. This refusal to commit to naturalism is less an ostrich-like reaction to the inevitable march of scientific progress than an allergic response to the guild-like arrogance of scientists. Philosophers typically approach the problem of knowledge with an open-minded attitude toward the means and ends of its pursuit. In contrast, the naturalistic privileging of certain disciplines, theories, and methods implies that most of these fundamental epistemological questions have been already resolved.

Like most philosophers, IDT proponents believe these matters are still worth keeping open, if only on grounds of human fallibility. Here one recalls the figure of the "skeptic" who always imagines an alternative explanation for any apparently straightforward phenomenon. Such an attitude is in marked contrast to what Thomas Kuhn (1970) famously called the "normal scientist" who presumes – until overwhelmed by unexplained phenomena – one explanatory framework, or "paradigm," within which she single-mindedly conducts her research. Much of the fire focused on IDT proceeds from these Kuhnian premises: that one is not entitled to offer a competing explanation for, say, the biochemical stability of a living cell, if the scientific establishment already provides a reasonably good one.

Philosophical open-mindedness toward the means and ends of knowledge has been typically expressed in a manner that is receptive – without being dogmatically wedded – to the multiple methods of science. Perhaps the leading American epistemologist of the last half-century, Roderick Chisholm (1982), drew a relevant distinction in his 1973 Aquinas Lecture at Marquette University in Milwaukee. On the one hand, some philosophers start from what we know and then ask how we know it. They presuppose that most of what we believe is true. On the other hand, some philosophers start from the view that we are always faced with conflicting knowledge claims and hence require a reliable decision procedure for resolving the conflict. They presuppose that much, though not all, of what we believe is false. In terms of the history of analytic philosophy – the dominant school in the English-speaking world – the former broadly capture the "ordinary language" philosophers, the latter the "logical positivist" philosophers. A key point to keep in mind in the ensuing discussion is that *neither* are naturalists – at least not without serious qualification.

The first class of philosophers grants that we know many things by many different means. Indeed, our ordinary ways of speaking may be treated as repositories of metaphysical insight. For example, the fact that we use two largely non-overlapping discourses to refer to our mental and physical lives suggests that there may be distinct realms of "mind" and "body," such that the former "transcends" the latter and hence may require a distinct mode of epistemic access. Philosophers who hold such a view tend to be metaphysical dualists or pluralists – that is, they believe that reality consists of more than one domain of being. In principle at least, they are open to supernaturalism, or at least their arguments readily lend themselves to support of supernaturalist metaphysics. One prominent philosopher whose work has been appropriated in this way is the Chomsky devotee, Jerry Fodor (1996), perhaps the most penetrating and persistent critic of the pan-Darwinist world-view, as promoted by Dennett (1995). Yet, while Fodor is reasonably classed anti-naturalistic, he is hardly anti-scientific. On the contrary, in the cognitive science community, he is read as a troubleshooter for problems in the design of experiments into mental phenomena.

Moreover, three distinguished mid-20th-century analytic philosophers, C. D. Broad (1887–1971), C. J. Ducasse (1881–1969), and H. H. Price (1899–1985), incorporated discussions of entities traditionally associated with the supernatural – paranormal phenomena and life after death – into their major epistemological treatises: Broad's *The Mind and Its Place in Nature* (1925), Ducasse's *Nature, Mind, and Death* (1951), and Price's *Thinking and Experience* (1953). None claimed to have demonstrated the existence of these traditionally supernatural phenomena, but they were careful not to let the issue be decided by methods inappropriate to the putative objects of inquiry simply because those were the methods used to decide other things scientists studied. Thus, all three philosophers devoted much time to criticizing attempts to terminate inquiry prematurely into the supernatural by scientists who proposed criteria of validity – as expressed in, say, experiments designed to detect the presence of "psychic" causation – that virtually guaranteed that the distinctiveness of the putative objects would *not* be registered.

The underlying principle shared by these philosophers is that the failure of orthodox scientific methods to register supernatural phenomena may just as much reflect a failure in the methodological imagination as the absence of the alleged phenomena. One should not reach for Ockham's Razor too quickly. Lest one suspect that such open-mindedness amounts to wishful thinking, it is worth observing that evolutionists have also benefited from a naturalistic version of it. After all, the historic failure of humans to engage in any but the most rudimentary communication with animals did not deter students of animal behavior from designing ever more subtle experiments that do not require animals to make sense of our talk in order for them to be reasonably attributed linguistic capacities. If there is a general tendency in the branch of neo-Darwinism called "evolutionary psychology," it is that the more sophisticated and clever the experiments designed by humans become, the smarter the animal subjects turn out to be. Their apparently meaningless utterances and behaviors appear more language-like. One might then equally countenance that smarter research designs will eventually enable us to attribute "intelligent design" to other entities – not simply at the sub-human but also at the sub-organic level – that do not *prima facie* possess those properties.

The second class of philosophers identified by Chisholm is more easily recognizable from debates concerning IDT and its creationist precursors. These are philosophers who presume that because we are always faced with multiple and often conflicting knowledge claims, we must settle on a method to decide the differences. In the philosophy of science, it gave rise to what is known as the "demarcation problem': how does one tell the difference between genuinely scientific theories and ones that merely pretend to be scientific, popularly known as "pseudo-science,"? Creationism was first judged (wanting) in these terms in the 1982 *McLean vs. Arkansas* case, when the presiding Judge William Overton invoked a general version of what Rudolf Carnap or Karl Popper would have recognized as "demarcation criteria:"

> A scientific theory must be tentative and always subject to revision or abandonment in light of facts that are consistent with, or falsify, the theory. A theory that is by its own terms dogmatic, absolutist, and never subject to revision is not a scientific theory. (Overton 1983: 62–3)

Most philosophers who have not participated in cases involving creationism or IDT regard such criteria as much too coarse-grained to capture the nature of scientific inquiry (e.g. Laudan 1982). In contrast, I happily accept them – as long as we are mindful of the spirit in which they were proposed. The logical positivists proposed demarcation criteria in order to provide a common medium for comparatively evaluating knowledge claims. Like a judicial proceeding, the criterion was to be specified in a manner neutral to the competing parties. Exactly what might count as a "neutral language" of evaluation led to endless wrangling amongst the positivists and their eventual abandonment of the demarcationist project. Nevertheless, they were agreed that the criteria could *not* simply amount to judging a newcomer theory in the terms of the incumbent. This was for two reasons: either the trial would be clearly biased to the incumbent, thereby inhibiting any truly novel breakthroughs in science, or the newcomer would be encouraged to recast its own knowledge claims in the incumbent's terms. The latter resulted in "pseudo-science," which the positivists found the more pressing problem in the 1920s and 1930s.

However, the original demarcationist concern with pseudo-science significantly differed from its reinvention in the 1980s during the creationist trials. This point is crucial to understanding my participation as an expert witness in *Kitzmiller*. The main examples of pseudo-science cited by the positivists were not drawn from the religious domain but rather Marxism and psychoanalysis. They were concerned mainly with attempts to extend legitimate scientific findings into domains where they pre-empted people's decision-making capacities, as in "Science shows that the revolution is coming," or "Science shows that your fate is sealed by age five." Karl Popper, who probably did the most to popularize demarcationism, believed that Marxism and psychoanalysis held valuable insights but that their proponents tended to overstate their significance in the public domain, where the evidence base was less secure – largely because the relevant knowledge claims could not be properly tested – yet action based on those claims could hold disastrous political consequences (Hacohen . 2000: ch. 5). In the next chapter, I shall return to this point, which prompted Popper's coinage of "historicism," a word that had previously existed in German but not English.

Were Popper and the positivists alive today, they would probably channel their demarcationist impulses not toward IDT or even creationism, but toward what Ruse (2005) has recently called "evolutionism," namely, the overextension of evolutionary arguments into the human realm, where their tendentiousness is masked by an epistemic "halo effect" from the high-quality research evolutionists have done on sub-human animals. Examples of such evolutionism appear daily in the public sphere to justify everything from the naturalness of rape to the genetic incapacity of women to do science. They are the stuff of such best-selling science popularizations as Pinker (2002). Of course, evolutionism dates back to Herbert Spencer, whose generalization of laissez-faire principles to the natural world preceded Darwin's *Origin of Species* by nearly a decade, and who continued to be a popular lecturer, especially in the United States, for nearly two decades after Darwin's death (Hofstadter 1955). More to the point, as we shall see below, the desire to drive a "wedge" between the empirical basis of evolution and its ideological conversion into an "ism" has given IDT forward political momentum (Johnson 2000).

4. America's Unique Legal and Political Culture

In the end, wrangles over the status of naturalism and supernaturalism in the history and philosophy of science must return to the requirements of high school science education. Here we need to keep in mind the original American context of this discussion. The interest in teaching science is shared by all sides to this dispute. Indeed, public opinion surveys consistently suggest that there would be a market for textbooks that creatively synthesize scientific and religious interests, given that two-thirds of Americans believe in both divine creation and evolution: Americans are clearly ready for a "re-enchantment" of science. (On the philosophical implications of this point, see Fuller 2006a: ch. 5.) Unfortunately things are never so simple.

By the standards of democracies in the developed world, the US displays a remarkable schizophrenia toward the production and distribution of knowledge. In the aftermath of World War II, the US National Science Foundation pioneered the peer review system that turned over research-funding decisions to elite disciplinary practitioners. Set against this relatively recent development is the traditional democratic attitude toward decisions about education enshrined in the US Constitution. Consequently, the US lacks a national education ministry capable of enforcing uniform curricula for primary and secondary schools. To be sure, the 50 states set academically respectable benchmarks for subject mastery, but exactly how they are met – by what textbooks and teaching methods – is typically entrusted to local school districts.

This policy is deliberate, reflecting the nation's origins in religious dissenters who, despite their relative prosperity, had been politically disenfranchised in their native Britain simply on grounds of belief. These dissenters were situated, both temporally and ideologically, between Calvin and Rousseau in the midst of the secularization of puritanical Christianity. For this tradition, a "commonwealth" is founded on a body of shared values and beliefs, with those unable to grant their consent expected, if not invited, to leave and form their own community of like-minded people. The settlement of the original British-American colonies exhibits just this pattern. It has given the US a historic reputation for pedagogical innovation and experimentation – instances of which have been both emulated and discarded, depending on their results.

In this context, the so-called "Establishment Clause" in the US Constitution's First Amendment – which expressly prohibits the establishment of a state religion like the Church of England – was designed to foster precisely the sort of religious self-organization manifested in the original settlers.

However, the introduction of state-based standards, combined with the consolidation of school districts early in the 20th century, began to shift the frame of reference for an appropriate curriculum. People from diverse class, religious, and ethnic backgrounds were increasingly thrown together – often united by little more than a common tax base – to hammer out mutually agreeable courses of study for children, who in turn may wish to leave their community upon adulthood. Indeed, the financial incentive for litigation – a consolidated school district's common tax base – has been crucial. The Supreme Court case normally credited with the current expanded interpretation of the Establishment Clause's scope, *Everson v. Board of Education* (1947), was brought by a disgruntled citizen who objected to his tax dollars being spent on subsidizing faith-based schools simply in the name of promoting free expression. The case set a precedent, so that nowadays all religious expression is effectively banned from publicly funded schools in the US, except as a formal object of study.

The Establishment Clause has thus evolved from a strategy to secularize the state to one that institutionalizes atheism: from being a mechanism to maximize the expression of religious opinion in public life, the Clause has come to justify the suppression of all such opinion, regardless of intellectual content or general pedagogical value. This point has generated a certain unease among legal scholars, who would much prefer to disqualify, say, "young Earth" creationism on the grounds that it peddles scientific falsehoods, rather than simply because it infuses the teaching of science with religious content (Carter 1993). Nevertheless, the US legal climate has made cases like *Kitzmiller* inevitable when some locals, themselves typically recent migrants, believe that their neighbors are trying to impose a religious orthodoxy in the curriculum.

Over the course of the 20th century, those who continued to champion "popular sovereignty" on educational matters, such as the populist politician William Jennings Bryan, a prosecutor at the celebrated "Monkey Trial" (about which more below), were increasingly portrayed as anti-intellectual bumpkins, not the rightful heirs of Thomas Jefferson (Hofstadter 1962: 125–7). In this quickly changing climate of opinion, science was presented as a reliable rallying point. The clear precedent was the rapid rise of Germany on the world scene shortly after its unification in 1870. Crucial to constructing a sense of solidarity amongst the disparate German principalities – historically divided between Catholics and Protestants, interspersed with a liberal sprinkling of Jews – was the introduction of scientific and technical training at all levels of the educational system, which by the eve of World War I had made Germany unsurpassed in pure science and second only to Britain in industrial production worldwide.

It is worth noting that, in practice, religious instruction has turned out to be offensive *only* when injected into the natural science curriculum. Supreme Court cases have not arisen over religiously inspired courses in the humanities or even the social sciences. What explains this rather specific sense of the problem of religion's status in the classroom? After successive waves of immigration from conflict-torn Southern and Eastern Europe, starting in the mid-19th century, it became common for self-styled "Progressive" thinkers to claim that natural science unites and religion divides in the great "melting pot" that the US had become. Indeed, some like John Dewey came close to arguing that the scientific method could function as a secular religion (White 1957). Although that vision was never quite realized, religion was nevertheless increasingly driven into the private sphere, on the assumption that while inherently non-scientific, religion was still, in an Orwellian sense, "compatible" with science, just as the "separate but equal" doctrine – prior to the Civil Rights Act of 1964 – suggested that blacks and whites could live together, as long as they did not interfere with each other: peace through apartheid. This curious assertion of compatibility – and the corresponding judgment that the defense was simply manufacturing an otherwise non-existent conflict between the scientific establishment and religious believers – was perhaps the most widely reported feature of the *Kitzmiller* ruling.

A larger contextualizing factor is the increasing role assumed by US state governments in the credentialing process in the early 20th century, through the issuing of the high school diploma (Hofstadter 1962: 125). The textbook under contention in the iconic "Monkey Trial," *State of Tennessee v. John T. Scopes* (1925), George William Hunter's *A Civic Biology*, had been in use in Tennessee school districts since 1909. Yet the Monkey Trial occurs only in 1925, once the state had established an educational authority laying down conditions under which a high school diploma may be given. In an earlier era, those offended by the majority views of the curriculum would have simply moved to another district, if not started their own school. What appeared from the state capital's perspective purely a measure to streamline the passage from the classroom to the workplace at the same time forced neighboring communities to confront and resolve their ideological differences in a common educational policy. In that spirit, the Tennessee state legislature passed the Butler Act in 1925 that forbade the teaching that humans descended from lower creatures, thereby explicitly denying the creation story in Genesis.

Given the subsequent course of American legal history, the targeting of Darwin as divisive turned out to be ironic. While the Butler Act clearly put Scopes in the wrong, the adverse media attention generated by the Monkey Trial gradually turned public opinion away from cutting-edge science to fundamentalist religion as the divisive element in an expanding heterogeneous society. The famed trial lawyer Clarence Darrow, who spearheaded Scopes's defense on behalf of the American Civil Liberties Union, took advantage of the case to showcase the potential of experts whose appeal to the "consensus of scientific opinion" could resolve matters of fact in difficult court cases. Thus, Darrow recruited paleontologists, eugenicists, and anthropologists. At the time, however, this gesture was taken more seriously by the media than the judge, who ruled virtually all of the experts irrelevant to the proceedings.

Here it is worth noting that in US law, the paradigm case for the scientist as "expert" is the applied psychologist whose knowledge straddles academic science and the liberal professions. Had such hybrids not been given early legal recognition, it is not clear that scientists, who may be highly regarded by peers for their research but lacking in clinical or forensic experience, would have been permitted to testify as experts. Indeed, Darrow himself had argued, nearly two decades earlier, that scientists typically had "more knowledge than judgment" and thus should not be assumed to be a fount of good sense for legal purposes (Golan 2004: 203). Lurking behind Darrow's early disparagement was a realization that scientific judgment is most secure in the laboratory or the seminar – that is, artificial environments designed to remove the very variables that often turn out to be most crucial in the law (cf. Fuller 1988: ch. 12). In terms of the philosophical discussion in the previous section, this realization – though abandoned by Darrow in the Monkey Trial – needs to be kept in mind in order to prevent the scientific theory of evolution from turning into a political ideology of evolutionism.

This is the context for considering the objections raised by Darwin's great opponent in the Monkey Trial, US populist politician and perennial Democratic Party presidential candidate, William Jennings Bryan, who served as expert witness for the prosecution. At the time, Darwinism was widely interpreted as a doctrine of biological determinism that justified the "survival of the fittest" mentality underwriting the most rapacious tendencies of big business and international militarism. Bryan himself had resigned as Secretary of State when President Woodrow Wilson plunged the US into World War I. He argued that the *prima facie* First Amendment claims of Scopes (as defended by the ACLU) were outweighed by the threat posed by the promulgation of a profoundly divisive speculative theory, as one might still regard the teaching of racialist theories today. While Bryan could be criticized for adopting an extremely "precautionary" approach to the teaching of Darwinism, it is worth noting that the textbook in use that raised his ire, Hunter's *A Civic Biology*, concluded with a lightly veiled defense of white supremacist eugenics (Kazin 2006: ch. 12). Indeed, "civic biology" was itself a euphemism for eugenics.

However, bolstered by the Hollywood film adaptation of the Monkey Trial, *Inherit the Wind*, which was nominated for four Academy Awards in 1961, Bryan and the Christian opposition to Darwinism have been usually presented from the standpoint of the trial's eyewitness raconteur, the acerbic journalist of *The Baltimore Sun*, H. L. Mencken, a stocky hardboiled man, played in the movie by a lithe and handsome Gene Kelly. Mencken, writing in the wake of America's geopolitical ascendancy at the end of World War I, portrayed Bryan and his followers as backward-looking isolationists. Perhaps the main cultural agenda being played out here was whether the United States should remain wedded to the Anglo-Saxon values on which the nation was founded or look to the rapidly advancing Germany for spiritual inspiration. Whereas Bryan warned early on against Germany's jingoism, Mencken unabashedly preferred Germany to Britain as the European harbinger of America's destiny.

Notwithstanding their British origins, Darwin and Spencer were taken up with greater enthusiasm, both scientifically and politically, by Germany, at least until the completion of the neo-

Darwinian synthesis, a decade or so after the Monkey Trial. Thus, Mencken and his *bien pensant* readers understood evolution not in relation to John Stuart Mill's liberalism but to Friedrich Nietzsche's longings for an *Übermensch*, an ideal that might come to be realized in the United States. It is also worth recalling that, by the 1920s, the largest American immigrant group for the previous two centuries had been Germanic (including Nordic), more than the combined ethnicities from the British (including Celtic). From that standpoint, Bryan was seen as representing a dying breed, still anchored in the nation's origins, more concerned with escaping a non-existent oppressor than assuming global leadership. (A good sense of just how close this Germanic sensibility came to dominating US politics may be seen in Philip Roth's counterfactual novel, *The Plot against America*, in which the celebrated aviator Charles Lindbergh defeats FDR in the 1940 presidential election, thereby aligning the US with Hitler in World War II.)

However, the Monkey Trial's legacy has come under increasing legal, political, and cultural scrutiny in a series of books, starting with *Darwin on Trial* (1991), by Phillip Johnson, a Berkeley law professor and former clerk to liberal Republican Chief Justice Earl Warren, whose landmark decision, *Brown v. Board of Education* (1954), led to the racial desegregation of US schools. After a sabbatical in the UK in the 1980s, Johnson became a born-again Christian and began to challenge the naturalism that Mencken and Hollywood had successfully associated with the scientific attitude in the popular imagination. Johnson hammered home the historically correct observation that naturalism is, strictly speaking, a metaphysical position with which many scientists and the scientific establishment have identified, especially since the ascendancy of Darwinism, but which is not necessary for an adequate – or perhaps even a fruitful – account of the means and ends of scientific inquiry.

Johnson was clearly worried that naturalism had come to function as a kind of loyalty oath to exclude the expression of unconventional "supernatural" views in science. His worst fears were realized in 1999 when the National Academy of Sciences (NAS) declared that the conduct of science is governed by a principle of "methodological naturalism," a point repeatedly made by the plaintiffs' experts in the *Kitzmiller* trial. It was as if contemporary science was so indefensible on its own merits that it required a philosophical fig leaf for protective cover. What surprises me most about this turn of events is that neither philosophers nor scientists have complained more about how NAS's endorsement of so-called methodological naturalism compromises the integrity of their respective disciplines.

"Methodological naturalism" has been promoted for several years by a think-tank based in Oakland, California, the wishfully named "National Center for Science Education," as a safeguard against IDT's supposed threat to the science curriculum. In truth, "methodological naturalism" is a pseudo-philosophy tailor-made to counteract a perceived pseudo-science: It is metaphysical naturalism pretending to be logical positivism. In other words, "methodological naturalism" is a neologism designed to capture two things at once that the history of the scientific method has tended to keep separate: the source of hypotheses and the conditions under which they are testable, what I earlier identified as the contexts of discovery and justification. This separation explains the studied neutrality that philosophers of the scientific method have tended to adopt toward "metaphysics," including *both* naturalism and supernaturalism: Neither metaphysical position offers a royal road to scientific validity, but both have had significant heuristic value. Not surprisingly, the scientific community's recent legitimatory appeals to methodological naturalism have appeared to sit uncomfortably even with philosophers who oppose IDT (e.g. Parsons 2005).

IDT's renunciation of the Bible's evidentiary privilege in scientific matters, combined with the NAS's resort to an ideologically exclusionary definition of science, impelled me to serve as a rebuttal witness in *Kitzmiller*. The campaign against IDT has largely been mounted on an appeal to fear and dogma, revealing a genuine lack of faith by defenders of the scientific orthodoxy in the lay democracy licensed to decide educational matters in the US political system.

To be sure, matters have not been helped by a liberal political culture in the US that has been long disenfranchised from the reins of power. Whereas what the historian Richard Hofstadter (1965) famously called "the paranoid style in American politics" has normally attached to anti-progressive trends, we have here a case of frustrated liberals adapting scare tactics reminiscent of Cold War anti-Communism, including the image of IDT as a front in "The Republican War on Science," to quote the title of a recent best-selling book (Mooney 2005). The Discovery Institute, the Seattle-based think-tank that has become the spiritual home of anti-Darwinism, is treated these days like the American Communist Party in the 1950s.

It is worth observing that the paranoid's fatal flaw turns not on strict accuracy but on proportionality of response. Thus, Senator Joseph McCarthy may have been correct that Communist sympathizers were employed by the US government in the 1950s, yet that bare fact did not justify his scare-mongering and bullying tactics, which only served to undermine the civil liberties his inquiries supposedly aimed to safeguard. Similarly, the fact that the anti-evolutionary scientific forces have been propelled by religious interests should not be sufficient to disqualify their theories from the classroom. By analogy with the McCarthy witch-hunts, more damage is done to the integrity of scientific inquiry than to its imagined opponents, who, lifted of the burden of having to defend the scientific establishment, are then free to present their challenge in the spirit of open dialogue (Campbell 1996). The phrase associated with such dialogue, "teaching the controversies," due to the literary critic Gerald Graff (1992), is unfortunately received with Orwellian derision by scientists living in contemporary America's confused political climate.

A telling detail is the intellectual background of Barbara Forrest, the Louisiana philosophy professor and expert witness for the plaintiffs in *Kitzmiller*, who managed to persuade Judge Jones that IDT was part of a vast conservative conspiracy to convert the US into a Christian polity, inscribed in the notorious "Wedge Document" (more about which in the next section), and thus should be deemed a religious social movement rather than an autonomous research program (cf. Forrest and Gross 2004). She is a leading scholar of the work of Sidney Hook, perhaps the student of John Dewey with the highest public profile. He was among the first New York intellectuals who already in the 1940s had made the pendulum swing from uncritical admiration to unrelenting suspicion of the Soviet Union. During this period, Hook was arguably the American philosopher most responsible for driving the émigré logical positivists to sublimate their leftist political instincts in what Rudolf Carnap called "the icy slopes of logic" (Reisch 2005). Thus, a movement that in interwar Europe had seen common cause between its own "unity of science" and Marxism's "unity of theory and practice" mutated, in the postwar years, into the depoliticized analytic philosophy establishment that still dominates the discipline in the Anglo-American world. Hook had been the intellectuals' McCarthy, and Forrest now plays Hook for today's McCarthys of secularism.

5. Conclusion: The Threat Posed by Intelligent Design

In my considered opinion, the defense did indeed have the weaker case in *Kitzmiller*, but equally the *Kitzmiller* ruling did an injustice to philosophy, politics, and ultimately to science. It reinforced the institutionalized atheism that has come to dominate US legal thinking about education at the expense of the heuristic role that belief in a designed "universe" has played in the advancement of science. I naively shared the defense team's hope that the presiding Judge John Jones might have ruled against the Dover school board, in light of its transparently religious motives, while taking a more adventurously open-minded attitude towards IDT, which pursues religious ends by scientific means. Although the Discovery Institute, the Seattle-based think-tank in the forefront of promoting IDT, condemned the *Kitzmiller* verdict for "judicial activism," I believe that Judge Jones simply took the path of least interpretive resistance to the precedents set by Establishment Clause rulings made in the recent past.

However inept in its self-understanding and self-presentation, IDT did not deserve to be dismissed outright. If IDT reads anything "literally," it is not the Bible but the history of biology, in which words implying intelligent agency, like "design" and "selection," have figured prominently – though with increasingly ironic inflection: e.g, "design without a designer" and "blind selection." IDT theorists rightly wonder why neo-Darwinists retain these theistic remnants, other than to avoid marking the clean break with anthropocentrism that is compelled by the logic of their theory (Menuge 2004: ch. 3). In the philosophically confused world of contemporary secular humanism, it is often overlooked that the mastery of "human nature," which supposedly enables us to "take control" of evolution, trades on differences between *Homo sapiens* (a coinage of the special creationist, Carolus Linnaeus) and other animals that are losing their salience with advances in the neo-Darwinian sciences of life. This point is vividly demonstrated – almost daily in the UK – in the trench warfare between animal rights activists and animal-based experimenters, both of whom claim Darwinian lineage. However, the activists, not the experimenters, are the Darwinian purists – and Peter Singer is their Calvin. It is no accident that IDT enthusiasts tend to hail from the more abstract and laboratory-based areas of science. They may turn out to be the animal experimenters' strongest allies in the long term.

Liberal US journalists who have followed the rise of IDT, such as *Washington Post* reporter Chris Mooney (2005), can only see the hand of the religious right at work. Moreover, their worst suspicions have been confirmed, as leading conservative commentators champion IDT's cause (e.g. Coulter 2006: chs 7–11). Yet there is more to this organized intellectual opposition to the neo-Darwinian paradigm in biology. Of course, some basic facts must be conceded at the outset: Yes, a line of descent can be drawn from high school science textbooks espousing biblical literalism to ones now espousing IDT. Yes, there is probably a strong desire, perhaps even a conspiracy, by religious fundamentalists to convert the US to a proper Christian polity, as epitomized in the notorious "Wedge Document" emanating from the Discovery Institute.

This document, whose authors remain anonymous, is a manifesto directed against America's rampant "materialist" culture, whose philosophical and scientific fronts are, respectively, naturalism and Darwinism. Specifically, materialism's repositioning of humans closer to animals than God is responsible for a variety of persistent global problems acknowledged across the political spectrum, ranging from economic deprivation to moral depravity, which are diagnosed in terms of "tolerance of the intolerable," the unflattering depiction of liberalism common to religious fundamentalists worldwide (Fuller 2006b: ch. 12).

Indeed, recalling the purely formal character of the demarcation criteria proposed by Ruse in *McLean v. Arkansas*, it is easy to imagine creationists cynically interpreting Judge Overton's endorsement as inviting minimal verbal adjustments to their texts to convey the impression that they have crossed the threshold into science. The events that precipitated *Kitzmiller* suggested just such a strategy, since the IDT textbook on offer – *Of Pandas and People* – began life as a "young Earth" creationist text (i.e. adhering to a literal reading of biblical chronology that results in a 6,000-year-old Earth), and the Dover school board members keen on it were more interested in IDT as a fig leaf for their creationist beliefs than in the details of the theory itself. As a result, the Discovery Institute refused to cooperate with the defense lawyers. To be sure, it also did not help that the person who turned out to be (after other defections) the "star witness" for IDT, the biochemist Michael Behe, is a contributor to the book's new edition.

The defense lawyers, perhaps anticipating these problems, confined my own testimony to the general matter of whether IDT counts as science. I was instructed not to speak with anyone on the Dover school board, nor read *Pandas*, the contents of which I only gleaned as testimony was given at the trial. While *Pandas* is not the book I would write to introduce IDT in a scientific light, it does imply that some forms of philosophical idealism and social constructivism might be considered versions of IDT. For example, in the *Kitzmiller* trial, the following quote from the textbook was cited as evidence that "intelligent design" is synonymous with "special creation": "Intelligent design means that various forms of life began abruptly through an intelligent agency with their distinctive features already intact: fish with fins and scales, birds with feathers, beaks, and wings, etc." (Davis and Kenyon 1993: 99–100). However, at this level of abstraction, it could equally serve as a somewhat reified account of how, in Kuhn's (1970) own words, "the world changes" in a paradigm-shift after a scientific revolution, since the paradigm-shifter acquires a new world-view as a whole, not in parts. More concretely, this account also captures the student's acquisition of the conceptual framework needed to address problems in her chosen science. The "intelligent agency" in this case amounts to the disciplinary instruction the student receives that enables a whole new domain of objects to come into view. Admittedly, all of this is small consolation, in light of *Kitzmiller*'s outcome.

But just how seriously should a theory's origins be taken as a mark of its validity? After all, every theory is born in an intellectual state of "original sin," actively promoted by special interests long before it is generally accepted as valid. It is therefore essential to monitor the theory's development – especially to see whether its mode of inquiry becomes dissociated from its origins. While IDT may appeal to those who believe in divine creation, its knowledge claims, and their evaluation, are couched in terms of laboratory experiments and probability theory that do not make any theistic references. Of course, this does not make the theory true but (so I believe) it makes it scientific.

Suppose we took the pulse of Darwinism in 1909, 50 years after the publication of *Origin of Species* but still a quarter-century before Mendelian genetics theory was generally accepted as providing the mechanism for an otherwise elusive process of natural selection. We would say that the theory's main backers were located outside the universities – even outside the emerging lab-based biological sciences. To be sure, the backers were not trivial players in the knowledge politics of the day. They included popular free-market intellectuals like Herbert Spencer, as well as many "captains of industry" whose self-understanding motivated their support of the fledgling fields of the social sciences, where "Social Darwinism" provided a powerful explanatory and legitimatory resource for the march of capitalism (Hofstadter 1955).

It is common for Darwinists to airbrush this bit of their history, which draws attention to the fact that as biologists struggled to identify the causal mechanisms responsible for the striking pattern of common descent and differential evolution that Darwin recorded in nature, congenial ideological currents – including eugenics and scientific racism – kept the theory in the public eye. Thus it is striking that, as part of the run-up to the bicentennial of Darwin's birth in 2009, the Darwin exhibition at the American Museum of Natural History in New York gives the misleading impression that any association between Darwin's theory and Thomas Malthus's anti-welfarist tract, *Essay on Population*, is purely coincidental. Yet Darwin himself acknowledged – and Darwin's admirers assumed – the profundity of Malthus's insight into the normal character of mass extinction, given the inevitability of resource scarcity. Contrary to the accounts usually given of Darwin's reception, what was provocative about *Origin of Species* was not the prospect that a theory of plant and animal species could also explain the human species, but the exact opposite: that a theory so obviously grounded in the explanatory framework of laissez-faire capitalism could be generalized across all of nature (Young 1985). Thus Darwin's toughest critics came from the physical and biological sciences, not the social sciences.

The ascent of Darwinism makes one wonder when the theory passed from being a well-evidenced ideology (say, like Marxism) to a properly testable science. Would it have passed the criteria used nowadays to disqualify creationism and IDT in, say, 1925, the time of the Monkey Trial? Probably not, since Darwinists still could not quite square their claims with cutting-edge genetics. However, it was equally clear that Darwinism enjoyed enormous support among self-styled progressive elements in American society who treated locally controlled school boards as the last bastions of intellectual backwardness. In this respect, the American Civil Liberties Union's showy intervention in *State of Tennessee v. John T. Scopes* employed a more successful version of the strategy now being carried out by the Discovery Institute and other organizational vehicles for realizing the "Wedge Document." Just as the ACLU helped to drive a wedge between the teaching of science and theology, the Discovery Institute would now drive a wedge between the teaching of science and the anti-theology prejudice euphemistically called "methodological naturalism."

I treat the two "wedges" as morally equivalent: Both can and should flourish under the aegis of American democracy without threatening its political foundations. As Darwinism slowly, fitfully, but finally made its way into high school and college classrooms, the theory was developed in new directions and integrated with new bodies of knowledge. As a result, Darwinism virtually – but of course never quite – distanced itself from its capitalist and racist roots, especially in cognate fields like socio-biology and evolutionary psychology (Fuller 2006b: ch. 6). I believe that a comparable fate awaits IDT in the coming decades, as it becomes an attractive option to those outside the orbit of religious fundamentalism and political conservatism. In chapter 2, I first raised the precedent of Herbert Simon's (1977) "sciences of the artificial" as a secular version of IDT's universal design perspective on nature.

Simon can be read as implying that a scientifically tractable way of thinking about "supernaturalism" is that the same form, end, or idea may be realized (or "instantiated") in radically different material structures. However, some structures may better suit their function than others. Here "designed" and "adapted" are, behaviorally speaking, little more than synonyms for "fit-for-purpose." Converting this general point into a program of theoretical and practical problems renders IDT scientific. Once we add IDT's metaphysical assumption that humans have been created in the image and likeness of God – or, less provocatively, that reality deeply resembles the structure of our minds – living things start to look like machines of somewhat better design than we normally make. We can treat these creatures as prototypes for technologies we might develop to enhance human dominion over nature: What function can this creature perform exceptionally well from which we can learn and then improve? Perhaps the most obvious historical example is the study of bird flight to develop aviation technology. In short, the biological sciences would become an advanced form of engineering – divine technology, if you will – corresponding roughly to fields currently known as "biomimetics" and "bionics," which draw very heavily and fruitfully from contemporary biology but without any theoretical commitment to the neo-Darwinian synthesis. In chapter

2, I mentioned Andrew McIntosh, the UK's leading intelligent design theorist, whose own scientific research conforms to this mold.

Besides insight into the integration of natural capacities and human interests, much money is potentially to be had by thinking of biology as a design-based discipline. This helps to explain why IDT has been so strongly backed by the Discovery Institute, which was founded by such technoscience sophisticates as Bruce Chapman and George Gilder, who have spent their careers exploring high-tech energy solutions and wealth-creation strategies. Put bluntly, they want to corner the market on "playing God" by both supporting the requisite technological innovations and laying down the moral ground rules for their use. Had liberal US journalists attended more to the continuities that have taken these young Rockefeller Republicans of the late 1960s to their current support of IDT, they might have also appreciated the Discovery Institute's reluctance to be too closely aligned with genuine biblical fundamentalists, as illustrated in the think-tank's withdrawal from support of the defense in *Kitzmiller*. Indeed, it should not have been too much to imagine that the Discovery Institute, whatever its intentions, is unlikely to succeed at spearheading a monolithic right-wing conspiracy, given that the fundamentalists who would be its foot soldiers simply want to read the Bible as a science book without having to grapple with the scientifically informed speculations of William Dembski or Michael Behe. Nevertheless, this division in the ranks, while marking a temporary political setback for IDT, is probably a sign of healthy intellectual development in the long term.

5

Life after Darwinism

1. Darwinism as Rhetorical Achievement

It is not too early to chart the intellectual course to the 22nd century. The 21st century may well mark a gradual disaffection with Darwinism, comparable to the 20th century's loss of support for Marxism. Indeed, Darwinism may be the last gasp of what Karl Popper (1957) called "historicism," a cluster of 19th-century theories that would grant history a measure of necessity comparable to what physics strives for, at least in terms of using our knowledge of the past, combined with general principles, to determine both what must and what cannot happen in the future. To be sure, Darwinism is both much broader and perhaps subtler in its claims than Marxism but the stamp of historicism remains. Just as Marxists held that history must proceed through a series of stages and that no revolution may happen before its time, Darwinists also depend on the image of evolution as a gradual exfoliation of species over a very extended timeframe that renders any rapid changes to the biosphere fraught with disaster. And just as Marx failed to anticipate the advent of the welfare state as obviating the need for revolution to alleviate the suffering of the masses, Darwin failed to anticipate that our knowledge of the microstructure of heredity would· become sufficiently secure to permit a genuine science of bioengineering.

It is worth noting that when Popper originally identified Marx and Darwin as sources of historicism in the mid-20th century, his critique was directed more at the likely consequences of believing these theories than at the theories' actual truth. It has been only in the past two decades that the truth of Marxist historicist claims have been generally seen as discredited, and the truth of Darwinist historicist claims are only now beginning to be questioned – for example, by the discovery that the same set of genes control similar functions in evolutionarily disparate creatures, which would seem to undercut the significance Darwinists have attached to the historical sequence of species, as iconically enshrined in successive layers of rock (Carroll 2005: 71–2).

Nevertheless, Popper's grouping together of Marxism and Darwinism as historicisms was justified not only by certain structural features they share as theories. It was also grounded in their shared prehistory in the late 18th-century Scottish anti-clericalism of David Hume, Adam Smith, and Adam Ferguson, all of whom conspicuously saw continuities between humans and other habit-forming creatures in ways that would not be out of place in today's evolutionary psychology research. Indeed, much of Marx and Darwin's common heritage carried into what we now regard as the early history of anthropology, not least through a figure like New York lawyer and amateur observer of the native American tribes, Lewis Henry Morgan, who in the 1870s popularized the tripartite sequence of savage/barbaric/civilized (now known as hunter-gatherer/agricultural/urban) forms of life, each tied to a specific logic of kinship ties (Harris 1968: chs 5–8). However, over the next 100 years, and despite their common universalist aspirations and Popper's own critical efforts, Marxism and

Darwinism came to be seen very differently within their respective fields of inquiry, the latter appearing very much more scientific than the former.

Nevertheless, by the end of the 21st century, neo-Darwinism may come to be seen as Karl Marx has taught us to view classical political economy, namely, an ideological gloss on some perfectly decent economic principles that conveys the misleading impression that whatever happens in the economy vindicates those principles. Here I envisage that a reader of neo-Darwinism in the year 2100 will smile knowingly at references to "natural selection" just as today's reader of neoclassical economics smiles knowingly at references to the "invisible hand." In both cases, the expression in quotes treats a certain account of history – of, say, the planet or Europe – as a natural evolutionary process. Darwinism will be no more synonymous with biology than capitalism is with economics.

Less pronounced in biology today than in political economy from the early 19th century onward are rival schools of thought that embed the same basic principles in alternative conceptual frameworks. Thus, the "law of supply and demand" was interpreted not merely as an emergent feature of extant economies but equally as a normative principle whose enforcement requires explicit anti-monopoly legislation. In this way, the "philosophical radicalism" of utilitarians like Jeremy Bentham and John Stuart Mill was distinguished from the more "laissez faire" political attitudes of Scottish Enlightenment figures like Smith and Hume (Fuller 2006b: chs 3–5). Of course, Marx provided a still more radical embedding of economics, interpreting the field's "laws" as alternative expressions of the commodification of labor power. To move biology in a similar direction would require a renewed focus on the central role that genetics plays in providing the "mechanism" for natural selection in the neo-Darwinian synthesis. Genetic potential is biology's answer to labor power (Fuller 2006b: 57–8). Perhaps unsurprisingly, both concepts have been dogged by unsavoury historical associations with "planning": the eugenic society and the command economy, respectively.

However, as it stands, the neo-Darwinian synthesis – what its admirers have historically sanitized as "modern evolutionary

theory" – is arguably science's most impressive rhetorical achievement (Ceccarelli 2001; Fuller 2006a: 174–9). Darwinism left the
20th century much stronger than it left the 19th. "Rhetorical"
here is meant with due respect, referring to all the biological
disciplines that have claimed intellectual legitimation from Charles
Darwin's *Origin of Species*. These range from historical and field-
based studies like paleontology and ecology and ethology to more
experimental sciences like genetics and molecular biology – as
well as the cross-cutting "interdisciplines" of socio-biology and
evolutionary psychology and applied fields like agronomy and
biotechnology.

A sense of the rhetorical character of the neo-Darwinian synthesis is evident from the phrase, "the origin of life," which
neo-Darwinists treat indifferently as referring to both the simplest
self-catalysing biochemical units and the historically earliest form
of life. But do we have good reason to believe that these two
senses of "life's origin" converge on the same referent? Why
should we think that the *oldest* form of life is the *simplest* form
possible? The answer is that neo-Darwinism assumes that molecular biology and paleontology provide alternative means of address-
ing the same fundamental questions. This is quite a remarkable
assumption, though it is easier to take for granted the more computer simulations are used to "test" and even "confirm" hypotheses about origins, as in artificial life research of the sort we saw
in the previous chapter was used in the *Kitzmiller* trial. The
assumption that "oldest" means "simplest" also stands behind
contemporary skepticism about the prospect that life on Earth
began in an already complex state as a form of alien migration – a
view entertained not only by proponents of "intelligent design
theory" (IDT) but also scientists like Francis Crick and Fred
Hoyle, who doubt that life could have evolved from scratch
exclusively through neo-Darwinian mechanisms, given the Earth's
estimated age.

To be sure, such objections recall the original problems that
physicists in Darwin's day like Kelvin had with his chronology –
that is, before the discovery of radioactive elements added orders
of magnitude to the Earth's age. And, of course, behind them lay
the biblically inspired image that life had enjoyed a more robust

existence prior to the onset of decline as punishment for Original Sin. However mythical that account may be, it nevertheless serves as a reminder that logical and temporal order are two separate matters that may or may not in fact converge: The past need not have been simple, and the future need not be complex.

Putting this matter in perspective requires considering the history of the social sciences. All of these disciplines pay lip service to wanting to discover the building blocks of social life. However, the social scientific versions of molecular biologists and paleontologists – experimental psychologists and field anthropologists – interpret the task rather differently. Psychologists experiment on members of their own society but in settings that abstract from the more obvious features of that society, leaving only those features that can be manipulated to elicit responses presumably indicative of the subjects' innate social dispositions. In contrast, anthropologists study pre-modern societies whose simplicity presumably reveals the mechanisms of social reproduction in a relatively unmediated form. Perhaps unsurprisingly, no current social science research program is dedicated to ensuring that psychology and anthropology reach agreement on the building blocks of social life. In fact, most social scientists doubt the prospect of such agreement, since the experimental method preferred by psychologists refuses to recognize any common history – that is, memory content – between the specific subjects, a crucial plot device in the narratives that flow from the anthropologists' preferred ethnographic method (Fuller 2006a: 174–9).

Here we see the rhetorical nature of neo-Darwinism's achievement. Biologists with rather different, perhaps even ideologically opposed, methods and theories have accepted the neo-Darwinian synthesis as an overarching explanatory framework. Social scientists in a comparable epistemic situation have not. We shall continue to draw reciprocal lessons from the histories of the social and the biological sciences in the rest of this chapter. At the most general level, biology's synthetic success inspires confidence, while sociology's synthetic failure cautions humility, especially in relation to the socio-biological mechanisms of "reproduction" and "inheritance." We shall see that issues of welfare, eugenics and ultimately racism were intertwined in the anglophone context in which the social and biological sciences disentangled their disciplinary identities in the first half of the 20th century. But first we need to consider a key feature of neo-Darwinism's rhetorical achievement, namely, the relatively loose hold that the theory has over actual disciplinary practices in biology.

2. The Artificiality of Darwinism's Dominance

Clearly exasperated by the contrariness of intelligent design theorists, journalist Chris Mooney exclaims at one point that Darwin's theory of evolution is "one of the most robust theories in the history of science" (Mooney 2005: 183). The claim is certainly familiar, but what exactly what does it mean and how might one determine its truth? Of course, Darwinism has had a persistent following for nearly 150 years, regardless of its evidential support. Moreover, Darwinism is philosophically "robust" insofar as it has caused philosophers to alter their definitions of science to accommodate a research program that clearly does not fit the mold of Newtonian mechanics. It is also true that most practicing biologists profess a belief in Darwinism, though the impact of that belief on day-to-day empirical research is harder to establish.

Published biological research makes surprisingly little reference to evolution or its principal Darwinian process, "natural selection." This point had been already made over a decade ago by the historian of 20th-century biomedical sciences, Nicolas Rasmussen (1994), who contended that neo-Darwinism was largely a philosophical cottage industry with little bearing on day-to-day biological research. I updated his finding. Based on the 1,273,417 articles from 1960 to 2006 indexed on the two main on-line biology databases on October 1, 2005, "evolution" and its variants appeared in the keywords and abstracts of 12 percent of articles, and "natural selection" in a mere 0.4 percent. Moreover, the incidence of the neo-Darwinian terms has been steadily *increasing* over the years – which is *not* what one would expect had Darwinism been empirically established before 1960 and is now taken as true without question. A spot check of the usage of "evolution" in these items reveals that the word is used to cover both observable, often experimentally induced, "microevolution" in the laboratory and more speculative inferences concerning "macroevolution" in the distant past based on the fossil record. The neo-Darwinian synthesis consists largely of an extended promissory note to the effect that these two senses of "evolution" are ultimately the same.

For a reality check, consider the Nobel Prizes awarded in the category associated with biology, "Physiology or Medicine." Nobel scientific committees are notoriously consensualist: They have freely rejected famous and accomplished nominees whose research cannot pass muster by peers from divergent national, theoretical, and ideological standpoints. Thus, Nobelists like T. H. Morgan, James Watson and Francis Crick, and Konrad Lorenz and Nikko Tinbergen have made discoveries that now bolster the neo-Darwinian synthesis, but they also less controversially contributed to the advancement of genetics or molecular biology or ethology. No one has ever won a Nobel Prize for having specifically contributed to the neo-Darwinian synthesis. Thus, absent from the roster of Nobelists are biologists who staked their careers on the synthesis: Ronald Fisher, Sewall Wright, J. B. S. Haldane, Ernst Mayr, George Gaylord Simpson, Theodosius Dobzhansky, W. D. Hamilton, George C. Williams, John Maynard Smith, E. O. Wilson, Richard Lewontin, Robert Trivers, and so on.

Nowadays the surest way for a US school board to enrage the scientific community is to pronounce that neo-Darwinism is "just a theory" that needs to be taught as such, perhaps alongside a competitor theory like IDT. Yet this also seems to have been the collective judgment of the Nobel Prize committee over the years. The point is obscured because popular accounts of biology, not least those written by biologists, give the impression that a new finding in, say, genetics is *ipso facto* another vindication of neo-Darwinism. On the contrary, neo-Darwinism could be abandoned tomorrow, and most research programs in genetics – and the other biological disciplines – would continue apace. Of course, this is not an argument against the validity of neo-Darwinian knowledge claims, but it does argue against their indispensability. The evidence could be just as well mobilized on behalf of another explanatory framework, say, one which takes design as more than simply a metaphorical feature of nature.

Insofar as the "neo-Darwinian synthesis" has been itself an object of inquiry, the inquirers have been philosophers of science and philosophically minded biologists who try to organize and reconcile research operating with theories and methods as varied as one might find in the social sciences. This is undoubtedly an intellectually interesting and challenging task. Nevertheless, it is hardly surprising that the logical positivists and their Popperian cousins classified neo-Darwinism as a "metaphysical research program." The phrase was meant to suggest that it inspires rather than justifies scientific work (Krimbas 2001). However, with the ascendancy of Kuhn's view that science is as scientists do, philosophers have come to expand their definition of science – which was traditionally anchored in Newtonian mechanics – to incorporate neo-Darwinism as exemplifying "another" kind of science (Rosenberg 2005).

This other kind of science does not pretend to universal cosmic reach, as it is based solely on the history of the Earth. Its metaphysical contours return us to the Aristotelian world-view that Newtonian mechanics had supposedly overthrown to establish the modern conception of science. In particular, neo-Darwinian science assumes that the Earth can be treated as "the privileged planet," a relatively closed system housing such unique developments as intelligence and even life itself not expected elsewhere in the universe (cf. Basalla 2006). Thus, neo-Darwinism's philosophical ascendancy appears to be a case of special pleading. At the very least it highlights philosophy's dual sense of guardianship vis-à-vis science: It functions both as *gatekeeper* for what counts as science and *reinforcer* of science's role as the exemplar of rational knowledge in society at large. It is in the latter context that philosophical appeals to Darwin's authority have mattered most – especially in secularizing research and education agendas worldwide.

However, even these appeals to Darwin are often little more than symbolic. For example, *Science* magazine declared 2005 the Year of Evolution, but what they meant by "evolution" relates rather loosely to what Darwin himself talked about. The magazine cited three developments: the sequencing of the chimpanzee genome, the mapping of the genetic variability of human diseases, and the emergence of a new species of bird. Only the last conforms to Darwin's own methods. Whereas he regarded natural selection as a process that occurred spontaneously in the wild and operated mainly on groups of organisms, today's breakthroughs in evolution occur mainly in the laboratory, often at the genomic or sub-genomic level, and are the product of explicit experimental interventions. That these two quite different senses of "natural selection" – sometimes distinguished as "macroevolution" and "microevolution" – are seen by paleontologists and geneticists alike as subsumed under the same "neo-Darwinian synthesis" draws attention, once again, to the theory's singular rhetorical achievement.

Although Darwinism starts in, say, 1860, and modern genetics is underway by, say, 1900, it is only in the period 1930–40 that the neo-Darwinian synthesis is forged, providing the covering theory for modern biological research. Indeed, by the fiftieth anniversary of the publication of *Origin of Species* in 1909, Darwin's theory itself was on the verge of extinction (Bowler 1988). The state of play in the biological sciences was then rather like that of the social sciences now. The ascendant science of experimental genetics and the declining one of natural history regarded each other with mutual suspicion and even contempt – not unlike how, say, psychologists and economists, on the one hand, today regard anthropologists and sociologists, on the other. "Natural selection" had become a vague catch-all for just-so adaptationist stories that failed to specify any consistent causal mechanism, let alone one that could account for apparent cases of discontinuity across species – what in the past would have been attributed to "special creation" but geneticists had come to explain as "mutation." Indeed, the leading theorist of heredity who supported Darwinism, Karl Pearson, was a statistician (or "biometrician") who did no experimental work himself, relying instead on probabilistic arguments based on "regression to the mean" to argue that extreme traits tend to "blend" over successive generations, unless they are specifically segregated – either to be cultivated as talented or culled as degenerate.

In striking contrast, experimental geneticists believed they had got hold of the mechanism of biological reproduction whose effects could be expressed in an algebraic equation (now known as the Hardy-Weinberg law) specifying the statistical distribution of traits in a stable population. Gregor Mendel, the Catholic monk who originated this idea without any concern for Darwinism, saw this mechanism as a mathematically rigorous version of special creationism that was programed into each generation of a given species (Olby 1997). In any case, geneticists were moving toward a universal – even "Newtonian" – science of biology that promised to render obsolete the concepts of "inheritance" and "heredity," both of which implied that each generation carries memory traces of its predecessors' lives in its own genetic material. Indeed, Wilhelm Johannsen, the Danish botanist credited with the coinage of "gene," objected to the residual animism of the British founder of genetics, William Bateson, who continued to think of genes as potential parts of organisms rather than as raw material that statistically combine to produce whole organisms. Johannsen's ruthlessly mechanistic approach to genetics paved the way for molecular biology to break down genes into ordered strings of amino acids (Moss 2003: 23–44).

In 1909 it would have been easy to imagine 20th-century biology as having developed along the lines of 20th-century social science, which lived a schizophrenic existence half-drawn to the natural sciences and half-drawn to the humanities. In such a counterfactual history, "biology" would have named a two-tracked – indeed "two cultured" – pursuit of only intermittent mutual interest. The history of genetics through molecular biology to biotechnology would be clear enough. It would have proceeded much as it actually did, but perhaps more quickly and with firmer links to physics and engineering. While Darwinists have been avid consumers of genetics, they have contributed relatively little to the field's technical progress. (Something similar may be said of sociologists and anthropologists who borrow from psychology and economics with little repayment of the compliment.) However, without Darwinism exercising normative constraint, there would now probably be greater tolerance and even adventurousness with regard to eugenics. Imagine a biological version of, on the one hand, the USSR's planned economy and, on the other, the USA's market economy. In fact, such biologizations actually existed but were never quite fully absorbed into their corresponding economic policies: cf. Lysenkoism (Roll-Hansen 2005) and Herrnstein and Murray (1994). In contrast, left to its own devices, Darwinism would have regressed to a virtually humanistic discipline preoccupied with wildlife conservation, as it increasingly appears in the writings of E. O. Wilson. It would be consistently critical of genetics-driven developments but largely ineffectual, due to its own ambivalent attitudes toward medical advances that "artificially" extend the human condition.

But of course, as a matter of historical fact, biology did not reproduce the two-tracked disciplinary structure that characterized 20th-century social science. Instead a "neo-Darwinian synthesis" was forged, the principal architect of which was Theodosius Dobzhansky (1900–75), whose *Genetics and the Origin of Species*, is the closest that biology has come to producing a text with the paradigmatic status of Newton's *Principia Mathematica* (Dobzhansky 1937; cf. Ceccarelli 2001: chs 2–3). At last geneticists and natural historians were presented with non-technical accounts of their mutual relevance in solving common research problems. Yet Dobzhansky's motivation for forging the synthesis was complex, and in retrospect is of a piece with mid-20th-century political realism.

On the one hand, Dobzhansky was a Ukrainian Orthodox Christian who believed that God used completely mechanical processes to design humans, who are endowed with the capacity to steer the course of evolution (Dobzhansky 1967). This is the context in which to understand one of the most capriciously contextualized quotes in the history of science: "Nothing in biology makes sense except in light of evolution" (Dobzhansky 1973b). The article, addressed to the US biology teachers, ends by referring to divine creation as among the things that only make sense in light of evolution. Like another founder of the neo-Darwinian synthesis, Julian Huxley, Dobzhansky was a devotee of the Jesuit paleontologist Pierre Teilhard de Chardin (1955). Here Dobzhansky is usefully read alongside other Cold War biological theorists like Conrad Waddington (Dickens 2000: ch.

6) and Gregory Bateson (1979), son of William, who believed that Lamarck could be redeemed once his account of evolution was transferred from the individual to the population level. In that case, eugenics could operate to promote more widely, by whatever means, traits that have proven advantageous to isolated individuals, so as to prevent default social tendencies from hardening into racial ones in the human environment.

On the other hand, the humility equally afforded by a theistic perspective explains Dobzhansky's reluctance to indulge these neo-Lamarckian impulses, unlike, say, his fellow synthesist, Julian Huxley. Interestingly, such humility also helps to make sense of Dobzhansky's lifelong membership and eventual chairmanship of the American Eugenics Society. As made clear in *Genetics and the Origin of Species*, Dobzhansky doubted the scientific competence of Darwinists – ranging from Britain's Fabian Socialists to Germany's National Socialists – who believed that all carriers of recessive genes should be sterilized, lest the entire gene pool be swamped with deficient offspring. Their draconian policies were based on a confusion of genetic potential and its realization in a particular generation. After all, Mendelian genetics teaches that recessive genes are a recurrent feature of the genetic make-up of all species, including those members who fail to manifest traits associated with the genes, and nature manages perfectly well to cope with them without the sort of global eugenic policies contemplated in 1937 by not only the Nazis and the Soviets, but also the Americans, the British, and the Scandinavians (Proctor 1988; Paul 1998; King 1999; Roll-Hansen 2005).

We see, then, the tempering role that a Darwinist sense of "nature" played vis-à-vis the wilder ambitions of eugenicists. Dobzhansky, himself a natural historian before working in T. H. Morgan's fruit-fly laboratory, turned to his advantage a traditional Darwinist objection to genetics, namely, that artificially manufactured organisms – say, the products of experimentally induced mutations – have a low survival rate in nature. (Consider here the fate of Dolly, the cloned sheep.) Thus, Dobzhansky promoted the idea that nature somehow (perhaps as the expression of divine intent) imposes a norm on the conduct of life, perhaps what Catholics still call "natural law," on which humans must tread carefully. To be sure, by today's standards, Dobzhansky remained a staunch eugenicist, calling for the compulsory sterilization of carriers of genes likely to produce disabled offspring. However, this was to be determined on a case-by-case basis, drawing on an array of empirical indicators. One of Dobzhansky's less noted legacies here was to have attenuated the racist connotations of eugenics – or what is nowadays called "genetic counselling" – by popularizing the neutral expression "genetic diversity" (Dobzhansky 1973a; for a contemporary defense and critique, respectively, see Cavalli-Sforza 2000; Fuller 2006b: ch. 8).

Thus, the main feat achieved by Dobzhansky and the other forgers of the neo-Darwinian synthesis was to persuade natural historians in Darwin's research tradition and laboratory geneticists in Mendel's research tradition of a strong analogy between their methodologically rather different pursuits. In time, macroevolution and microevolution came to be understood as "evolution" in exactly the same sense. A comparable development for some aspiring covering theory of the social sciences would be to convince, say, historical anthropologists and experimental economists that the "markets" unearthed in the ancient world and constructed in the laboratory are to be explained by the same mechanisms, which the latter research environment reveals in their pure form. Included among the obstacles to such a synthesis being forged in the social sciences is the perceived incommensurability between "qualitative" and "quantitative" research methods. One consequence of the neo-Darwinian synthesis was to break down these Aristotelian hang-ups, which had also existed in biology, permitting both methods to migrate across the micro–macro divide with fruitful research results.

It follows that the construction of neo-Darwinian synthesis has much to teach the social sciences, progress in which has been retarded by the sort of "metaphysical" suspicions that neo-Darwinism gladly suspends. Nevertheless, there remain fault lines in the synthesis, which occasionally surface, especially in the popular science literature, where the underlying assumptions and projected implications of empirical knowledge claims are discussed more openly than is normally permitted in the consensus-driven world of peer review. These fault lines may be uncovered by asking two kinds of biologist, a field scientist and a lab scientist, what the theory of "evolution by natural selection" is supposed to be *about*.

The lab scientist would probably say that it is a model of potentially universal scope, with the actual history of life on Earth as merely one – and perhaps not even the most important – confirmation of the theory. She would probably not lose too much sleep, were she to learn that natural selection proves insufficient to the task of explaining the entire history of life on Earth because the model still applies in all sorts of smaller and maybe even larger domains (e.g. Lee Smolin's theory of cosmological selection). In contrast, the field scientist would turn the tables and say quite plainly that the theory of natural selection is exactly about the actual history of life on Earth, and that the fate of the theory rests precisely on the extent to which it explains the patterns that Darwin and subsequent natural historians have found. Everything else is merely a metaphorical extension of the original theory.

This is quite a serious difference of opinion in how one defines a theory's referent. Perhaps, then, neo-Darwinism is so "robust" because it is so strategically vague – or should I say, "adaptive"! Nevertheless, the fault lines are periodically revealed. The late Stephen Jay Gould, whose expertise was closest to Darwin's own (not least in his general ignorance and disdain of lab-based science), fits my "field scientist" to a tee. Not surprisingly, then, as the evidence from extant and extinct creatures suggested the insufficiency of natural selection as an overarching explanation for the actual history of life on Earth, he became pan-Darwinism's fiercest critic. Many neo-Darwinists have not only decried Gould's perceived defection from the fold but have more harshly criticized intelligent design theorists for trying to get some mileage from Gould's apostasy (Wright 1999; Woodward 2003).

But all of this seems to suggest that the neo-Darwinists have proprietary rights over the entire history of biology. Yet, neo-Darwinism's own pivotal mechanism – what is now called "Mendelian genetics" – was contributed by people who held the counter-Darwinian assumption that every member of a species, regardless of its generational history, is programed with a reproductive propensity. That assumption is a legacy of special creationism, a research tradition in natural history that connects the devout Christians, Linnaeus, Cuvier, and Mendel. To be sure, many of its elements have been subsumed by the neo-Darwinian synthesis. But why can't intelligent design theorists reclaim this subsumed tradition as their own to develop the biological sciences in a different direction? In that case, Gould is rightly invoked as an ally – if only in a backhanded way – because he stuck to Darwin's original formulation of evolutionary theory and found it empirically wanting, whereas neo-Darwinists have shifted the goalpost to make it seem as though the theory's validity rests more on evidence from the lab than the field.

In short, intelligent design theorists treat what evolutionists regard as a *broadening* of evolutionary theory's scope, which corresponds to the ascendancy of lab-based research, as involving a *thinning* of the theory's content. This point first struck me as an expert witness for the defence in *Kitzmiller*. One expert witness called by the plaintiffs, whom Mooney also quotes as a source, was the philosopher Robert Pennock. He enthused under oath about an "artificial life" computer program that he and some colleagues at Michigan State University had recently written up for *Nature* (Lenski et al. 2003). To the unprejudiced observer, the program looks like a strategy for generating computer viruses without the user's intervention, albeit within parameters that approximate the combinatorial tendencies of DNA. Yet Pennock claimed that this program "instantiated" evolution by natural selection. The metaphysically freighted "instantiated," much favored by artificial-life researchers, renovates the old theological idea that essentially the same idea can be materialized in radically different ways. This idea was originally used to justify God's Trinitarian nature. Too bad, under cross examination, Pennock wasn't asked whether he thought his program *added* to neo-Darwinism's success at explaining the history of life on Earth – or merely *substituted* for it. So much for neo-Darwinism's falsifiability!

Evolutionists have been allowed to hedge their bets in this fashion because, prior to the neo-Darwinian synthesis, there had been no "robust" theory of the biological sciences as a whole. Biology was a scientific free zone, which is easily documented by noting the non-university locations of many of its historic practitioners. Under the circumstances, it is easy to understand – but no less unfortunate – that a journalist like Mooney could come to make a simple equation between neo-Darwinism and biological science as such. This leads him to suspect, unjustly I believe, that IDT, alternatively cast as pseudoscience and anti-science, is conspiring to replace neo-Darwinism wholesale – with some sort of biblical fundamentalism.

At most, intelligent design theorists are guilty of opportunism here, exploiting substantial differences of opinion already present in the neo-Darwinian ranks, which the parties themselves think should be discussed in peer-reviewed publications rather than in the media, courtrooms, and classrooms. Thus, intelligent design theorists typically accept exactly the sort of microevolution evidence which led *Science* to declare 2005 as the Year of Evolution. But that's because "evolution by natural selection" in these cases has been intelligently designed, namely, by the human researchers responsible for setting up the relevant experimental conditions. But what would allow natural selection to work so decisively in nature, without the presence of humans? That was the question that really interested Darwin – and Gould. It drove the analogy between "natural selection" and "artificial selection," which of course refers to the human breeding of animals. At this point, IDT dissents from the neo-Darwinian orthodoxy and refuses to accept macroevolution as the final word.

Is there more genuine intellectual disagreement between those who do and do not take Darwin to have laid the incontrovertible foundations of modern biological science than among those who grant Darwin such an exalted status? In terms of scientific personalities, is the difference between Dembski and Dawkins greater than between Gould and Dawkins? In the current scientific cold war, the answer would seem to be obviously "yes." However, the answer becomes less clear, once the neo-Darwinian synthesis is seen as a relatively baggy theoretical construction, parts of which can be reasonably believed without believing the whole. It is telling that when 67 national academies of science published a joint statement in June 2006 on the need to teach evolution as fact, the only beliefs clearly opposed by the agreed propositions were those of six-day creationists. For example, no appeal was made to natural selection at all, let alone as the primary explanation for evolutionary change. This was probably because the academies found natural selection a more controversial thesis than that the Earth has existed for 4.5 billion years.

The airing of such thinly veiled disagreements is vital for the elaboration and development of biological science. However, this spirit is lost if the neo-Darwinian synthesis must be accepted as a whole, perhaps even as a necessary condition for being taken seriously by the scientific establishment. The result, which captures the current state of play in the US, distorts the intellectual horizon by minimizing differences amongst Darwinists, while maximizing differences between them and their dreaded religiously inspired opponents. A recent high-profile expression of this distortion is *Intelligent Thought: Science versus the Intelligent Design Movement* (Brockman 2006a), whose editor, literary agent to the scientific star John Brockman, sent a copy to every member of the US Congress. Each copy enclosed a letter (Brockman 2006b), written to whip up a Joseph McCarthy-like frenzy that no less than American economic competitiveness and national security are at risk from the promotion of IDT. Fortunately this maneuver has had no discernible effect on the members of Congress. Meanwhile, across the Atlantic, at least two major British universities (Leeds and Leicester) began to include some lectures on the controversies surrounding evolution and intelligent design in their introductory zoology classes, in autumn 2006. While the UK Secretary of State for Education has so far banned the use of IDT materials in publicly funded secondary schools, the private schools using them have consistently performed above average on national science exams.

3. Life Spared of Darwin: The Case of Sociology

Patrick Geddes and Victor Branford are nowadays largely forgotten, yet they are not unfairly regarded as the Marx and Engels of British sociology's would-be Darwinization. Both Geddes and

Branford trained in natural history, the one visionary and undisciplined (like Marx); the other moneyed and more strategic (like Engels). Together they established Britain's first self-consciously "sociological" association and journal in the first years of the 20th century. To appreciate the sort of science they thought they had founded, consider this statement by Branford in 1904, from the inaugural meeting of the Sociological Society:

> as affecting the genesis of Sociology, the main features of the century were, in the first place, the creation of the Biological Sciences as definite systems of study, and in the second place the growth of the conception of a Science of History. (Branford, quoted in Dahrendorf 1995: 97)

As Ralf Dahrendorf (1995: 98) astutely observes, Branford (and Geddes) basically saw sociology as the marriage of eugenics and evolution.

Early British sociology is usually explained in terms of three distinct schools of eugenics (Galton et al.), environmentalism (Geddes et al.), and the institutionally dominant ethical socialism (Hobhouse et al.), all of which enjoyed some sympathy with Sidney and Beatrice Webb, the Fabian founders of the London School of Economics (LSE), where social science first took academic root in Britain. However, here a little historiographical reflexivity is in order. The difference between the so-called eugenicists and environmentalists in the first decade of the 20th century was really not as great as it might have appeared to someone writing in the 1960s, when this tripartite structure was first advanced (Halliday 1968).

For obvious political reasons, historians of both the biological and the social sciences in the generation following World War II strove hard to circumscribe the impact of hereditarian doctrines on modern social thought. Many of the relevant rhetorical innovations are still used uncritically today. In matters relating to genetics, it became easier – or at least more permissible – to tell when trained geneticists were "misusing" their skills than when scientists in other specialities were misusing theirs. At a more theoretically substantive level, two sorts of arguments were increasingly used to drive genetics out of the social imaginary. First, biological evolution has occurred over such an extended timeframe as to exclude its impact on social change in historical time. Second, the cells responsible for reproducing life are sufficiently segregated from the cells constituting a living organism (aka the Weismann doctrine) that nothing experienced in an organism's lifetime could be directly inherited by its offspring. Both arguments still manage to keep sociology at arm's length from genetics – just as long as we ignore the prospect of successful targeted technological intervention in the human genome itself. However, what was reasonable to ignore in the 20th century is becoming much less so in the 21st.

Indeed, we have returned to a point in history rather like the one that the founders of sociology (in all the major national traditions) faced almost exactly 100 years ago: namely, negotiating the boundary – if there is to be one – between something called "biology" and "sociology." Fields with names like "socio-biology" "evolutionary psychology," "behavioral genetics," "cognitive neuroscience," and "cognitive anthropology" increasingly draw students today who in the immediate postwar generation would have stayed within the established disciplinary boundaries of the social and the biological sciences. For their part, Geddes and Branford certainly addressed concerns not traditionally associated with sociology, but are now being rediscovered, especially the role of the environment, both natural and built, on the movement of peoples, and the constitution of social life more generally (Studholme 2007; Scott and Husbands 2007). But a disservice would be done to these neglected would-be founding fathers and ourselves if such historical observations were made simply in the spirit of adding an annex to a well-designed disciplinary edifice. Geddes and Branford really wanted to found sociology on a radically different basis. The basis is not one that appeals to me, but we do live in a time when the proposal can be taken seriously again. With that in mind, let us return to the quoted passage from Branford's 1904 address.

By the "biological sciences," Branford simply meant eugenics, the field that, in retrospect, perhaps did the most to popularize the power and versatility of statistics as a scientific mode of reasoning. Of course, this was not the primary aim of Francis Galton and his followers, notably Karl Pearson, who mistakenly believed that statistics *as such* had isolated the laws of heredity, which in turn carried substantive implications for social policy. In particular, the "regression to the mean" principle, associated with successively larger samples of a population, was tied to a blending theory of genetic transmission whereby, without segregation, two parents displaying polar traits will produce offspring that blended those traits, which after successive generations would lead to a virtual elimination of the founding parental differences. If the trait in question is intelligence, it would seem to imply that, unless specific policy measures are taken, society is ultimately swamped with mediocrity.

Now, by the "science of history," Branford meant evolution, which should be understood as a scientific approach to history indifferent to the social–biological distinction. Here it is worth recalling that today's sense of the "evolutionary synthesis" − that is, the formal establishment of Mendelian genetics as the causal mechanism responsible for the spontaneous variation in offspring on which Darwinian natural selection operates − was canonized only in the 1940s. Before that time, a variety of possible genetic mechanisms were in play. These included not only the Mendelian view, which treats an individual's expressed traits as only part of the individual's genetic potential, but also the blending view favored by Galton and the original Darwinists, which equates an individual's genetic potential with its family history. In addition, there were of course the followers of Lamarck, who postulated a genetic memory of the species (or race) that could be affected by an individual's experience. Taken together, along with Darwin's own widely read account of human evolution, *The Descent of Man* (1870), it was quite common to see Spencer and Darwin as "evolutionists" in much the same sense.

My point here is that had Branford's preferred candidate, the botanist Geddes, rather than the philosopher Leonard Hobhouse, been awarded the UK's flagship chair in sociology at the LSE in 1907, a substantial institutional precedent would have been set for "sociology" to be the name of a covering science under which the disciplines we now call "sociology" *and* "biology" would

have been subsumed. The significance of this counterfactual prospect should not be underestimated, especially given Britain's
overall geopolitical prominence at the time – albeit second in
science to Germany. It could have given a head start of at least
seven decades to the ascendant interdisciplinary research programs
mentioned above that nowadays blur the social–biological distinction. After all, E. O. Wilson's *Sociobiology: The New Synthesis* was
published only in 1975.

Although Geddes and Branford tended to be opportunistic in
their appeals to Comtean positivism, they took seriously Comte's
view that sociology would usher in a reflexive understanding of
the history of science. For Comte, this amounted to sociology's
rationalization of science's penultimate stage, biology, the field
that Lamarck had canonized in Comte's schooldays. However, in
the wake of Darwin's intellectual revolution, the horizons of the
field that Geddes and Branford called "biology" were hardly those
of Lamarck and Comte. The two Frenchmen held that lower
organisms literally strove to become higher organisms, specifically
humans, who at some point in the future would be Earth's sole
denizens. In contrast, the two British sociologists shifted the focus
from self-improving organisms to environments that facilitate the
working of natural selection, which is presumed to have, so to
speak, a mind of its own – that is, not necessarily oriented to the
endless proliferation of humans on the planet. I shall return to
this point below, for while Geddes and Branford certainly promoted an *ecocentric* sociology, it is by no means clear that it was
also *anthropocentric*.

Here it is worth observing that first-generation supporters of
Darwin tended to gravitate toward one of two polar views concerning the normative implications of natural selection for the
human condition. Some followed Spencer (and probably Darwin
himself) in believing that nature holds the ultimate trump card in
the prospects for collective human progress. Very much in the
spirit advocated today by, say, Pinker (2002), natural selection
should remind policymakers of the limits on our ability to improve
fate through legislation and artifice. Extending the lives of the
poor through welfare policies simply wastes resources and post-
pones the inevitable extinction of the unfit. In contrast, others
followed Thomas Henry Huxley, himself a trained surgeon, who

treated one person's death as a collective human failure. Huxley held that precisely what distinguishes humans from other animals is our systematic resistance and occasional reversal of natural selection through the hallmarks of modern civilization, especially the rule of law and medical science. Whatever one makes of these two responses today (and I squarely side with Huxley: Fuller 2006b: 141–3), in their day Spencer was seen as a pacifist and sometimes a multiculturalist, whereas Huxley was regarded as an ethnocentric imperialist.

Like many second-generation Darwinists, Geddes and Branford tried to finesse the normative schism that had opened up between Spencer and Huxley by the 1890s. In practice, this meant a realignment of Huxley's position, which was widely seen as conceding too much to Christian sensibilities, to bring it closer to Spencer's uncompromising secularism. More specifically, they granted Huxley's point that civilization had radically restructured the selection environment. Modern public hygiene and medical innovations had reduced infant mortality rates, resulting in unprecedented population growth, which, when combined with improvements in communication and transportation, allowed for an equally unparalleled movement of goods and people. However, whereas Huxley saw these developments as unmitigated progress, Geddes and Branford believed that they were destabilizing the environment, inhibiting or perhaps even perverting the operation of natural selection.

Even before Huxley formally broke with Spencer in his famous 1893 Romanes Lecture, "Evolution and Ethics," Geddes was already voicing concerns about the unintended consequences of civilization on the planet. In "Variation and Selection," written for the ninth edition (1888) of the *Encyclopaedia Britannica*, the most authoritative reference work in the English language, Geddes alerted readers to the prospect of "retrograde selection" and even "degeneration" of *Homo sapiens* because artificially enhanced environments enabled more unfit people to survive longer. In this respect, his concerns were very similar to that of the eugenicists.

However, what makes Geddes a more plausible candidate than, say, Galton or even Pearson for being a lost founder of a biologized sociology is that his definition of fitness had a much stronger social inflection. The eugenicists were ultimately inspired by Plato's ideal of an intellectual aristocracy, which led them to update his project of preventing the best from being debased by intermingling with the rest. They took fitness in human populations to be a property of the intellect possessed by individuals exclusive of their social relations. In contrast, for Geddes, fitness lay in the capacity for cooperative behavior, which requires everyone to pull their own weight so as to complement the efforts of others, from which all then may benefit. From that standpoint, those unwilling or unable to make an adequate contribution to the collective project are a drag on the entire ecosystem: They consume resources without producing anything in return. Those who shared Marx's fascination with Darwin, like August Bebel, a founder of the German Social Democratic Party, envisaged the rich as just such parasites, though the image soon migrated, courtesy of sociologists like Werner Sombart, to cover the Jews, many of whom could be counted among the rich (Fuller 2006a: 136; Fuller 2006b: 186–7). I shall explore this sinister, but by no means idiosyncratic, interpretation of Darwin in what follows.

As implied by its appearance in the *Britannica*, Geddes' interpretation of Darwin's theory was by no means eccentric. For example, human geographers, notably the anarchist theorist Peter Kropotkin, subscribed to such a view. Interestingly, so too did the renowned philosopher Bernard Bosanquet, who was trained in the same Oxford idealism as Geddes' LSE nemesis, Hobhouse. Bosanquet approvingly read Geddes as pointing to a higher-order version of natural selection that went beyond the materially fit to identify the spiritually fit – that is, those whose character renders them fit for membership in civil society (Bosanquet 1899: 25). By the end of World War I, Hobhouse would challenge this conservative drift in the neo-Hegelian heritage he shared with Bosanquet for its neglect of material conditions that remained very real for the poor and obviated any easy judgments of their fitness for civil society, especially (as Bosanquet suggested) on the basis of their spontaneous sense of personal initiative (Collini 1976). At this point, a line had been drawn in the sand between "Right" (Bosanquet) and "Left" (Hobhouse) heirs to the neo-

Hegelian movement that originated with Thomas Hill Green, comparable – albeit in more politically muted terms – to the line drawn between "Right" and "Left" followers of Hegel in Germany in the 1840s.

Geddes, being of a more practical disposition, did not settle for the philosopher's prerogative of patronizing or demonizing the surplus unfit: he tried to do something about them. Geddes early expressed disappointment that Spencer would not support Galton's proposals for the artificial sterilization of the surplus unfit. Geddes found this a more humane and scientific solution than Spencer's preferred policy of letting nature engage in its own slow war of attrition against overpopulation (La Vergata 2000: 195). However, more emblematic of Geddes' approach was the design of environments that "recover the best ideals of the past and reinstate them in the fresh light of evolution" (Geddes, quoted in Kent 1981: 92). The implied distinction here, between "ancestral" and "functional" environments, was one common to racialist forms of social science at the time, not least the German racial hygiene movement that pre-dated Hitler's rise by nearly a half-century and found inspiration in Geddes' "garden cities" approach to town planning (Schubert 2004; cf. Fuller 2006b: 187–95).

What nowadays is given the sanitized name of "environmental sociology," but which in the 20th century has been also called "human ecology" and "racial hygiene," is basically a scientifically updated version of the original "blood and soil" approach to culture, which held that the right built environment could awaken a collective memory in a people who have inhabited a region for many generations, thereby consolidating them and improving their chances for survival (Weikart 2005). Although the theoretical basis of this approach in genetics always remained hazy, there was a clear Darwinian precedent in the definition of race as a proto-species that results from the division of a common population into geographically segregated units that engage in mutually exclusive patterns of reproduction over many generations.

Recall the sub-title of Darwin's *Origin of Species: The Preservation of Favoured Races in the Struggle for Life.* The presumed difference between race and species is one of degree, not kind. Two populations descending from a common ancestor belong to different species if interbreeding fails to produce fertile offspring. This definition, carried over into the neo-Darwinian synthesis by the German-trained Harvard taxonomist Ernst Mayr (1942), reveals the futility of any easy enrollment of "biology" into contemporary "realist" approaches to social theory. Darwinists have always operated with what can be only called a "constructivist," perhaps even "performative," attitude toward species: namely, you are a member of the species that you are capable of sexually constructing (Fuller 2006b: 199–200). The interesting question then is what to do about specifically *racial* differences, which might either amplify into species differences after several more generations of segregated reproduction or disappear altogether through interracial breeding. Here many Darwinists have sought normative guidance from nature in what is nowadays called "biodiversity," which valorizes the inclusion of the maximum number of species – even if that means segregating (so as not to eliminate racial differences) and culling (so as to allow room for emergent species) populations.

Geddes and Branford should be understood as major British contributors to this extension of Darwinian reasoning into the human realm. It helps to explain their preference for regions over states as units of social inquiry and policy. The former support the spontaneous heterogeneity of nature, whereas the latter homogenize regional differences by enforcing common standards, which ultimately result in conflict, both within and between states. For Geddes, Branford, and the intellectuals attracted to their views in the trendy modernist magazine, the *New Age* (the source of today's expression, including its association with "holism"), World War I and, after it, the League of Nations were testimonies to the futility of state-based forms of social organization that promote an artificial internationalism, be it based on capitalism or socialism (Collini 2006: 91–101). This is the context in which we should understand their call for a "third alternative" focused on the university as an institutional magnet to consolidate regional identity (Geddes and Branford 1919).

Although, as Scott and Husbands (2007) show, the University of Michigan provided the model, Geddes realized the ideal at the

Hebrew University of Jerusalem, which was among his projects (including the city plan for Tel-Aviv) commissioned with the help of the British Foreign Secretary and former Sociological Society member, Arthur Balfour, whose 1917 Declaration on Palestine first granted political recognition to a Jewish homeland. As Geddes and Branford would have it, Hebrew University attracted many distinguished visiting professors, including Freud and Einstein, whose *de facto* affirmation of their Jewish identity inspired others to study there, often leading to permanent settlement in Palestine.

There are two general ways to think about the *empirical* basis of the future. One is as the inductive continuation of the immediate past. In that case, radical transformation comes about only through an incremental extension of the same process over time until a change in degree metamorphoses into a change in kind. Communist utopias predicated on capitalism's self-destruction belong to this category. The other empirical basis is as the recurrence of an earlier moment that was open to alternative futures. The sharp nature–nurture dichotomy that operated as an intellectual demilitarized zone between biology and sociology in the postwar era is now dissolving, returning us to the open borders between the two fields that existed at the start of the recently departed century. The reader should think about the arguments presented here in that spirit.

Reviving Geddes and Branford today would mean seriously reconnecting with geography and biology as sources of social identity. Notwithstanding the current popularity for something called "regionalism," the term remains vacuous (or negatively defined as "anti-statism") without an account of how and why people are tied to specific physical spaces. People like Geddes and Branford thought the world would be at peace once everyone had found a place they could call home (*Heimat* in German). They based this claim more on the current thinking about the physical and biological basis of social life than on any sustained study of the beliefs and desires of the people concerned (Kent 1981: 91–3). It celebrated multiracial diversity, with each race in its proper place, and diagnosed the global instability caused by both finance capitalism and international socialism in terms of the

uprooted nature of human interaction that such movements promoted. The idea of a Jewish homeland was given forward
momentum after World War I, partly because Jews were seen as
a people whose existential dislocation was expressed precisely
through their involvement in these movements.

Because both Branford and Geddes died shortly before Hitler
came to power, their memory is spared the taint of an early
infatuation with Nazism's avowed regionalism (unlike, say, fellow
modernist visionaries like Wyndham Lewis and Lewis Mumford).
Their thought thus provides valuable insight into a path not
taken, one that followed through the original complementarity
of Nazism and Zionism, whereby the latter is proposed as a
"cure" for the "disease" diagnosed by the former – but without
postulating the need for Hitler's "final solution." Moreover, I
believe that the Holocaust, now so intimately linked with the
fate of Nazism, was in fact an extreme policy even within Nazism's
own terms, reflecting more Hitler's circumstances and state of
mind than the logic of an ideology that could just as easily have
pursued policies that would have retained Branford and Geddes'
interest, if not outright endorsement. (I pursue this point, as these
ideas relate to the roots of today's "Green" political sensibilities,
in Fuller 2006b: ch. 14.) Nevertheless, all told, the LSE was
probably right to have chosen Hobhouse over Geddes for the
first UK sociology chair, which in turn hampered efforts to bring
an emerging welfarist sensibility in line with Darwinist biology.

4. Conclusion: Alternative Scenarios for Life after Darwin

The history of science in the first half of the 20th century may
be read as a field of competing synthetic projects that would have
brought together the social and biological disciplines, in various
proportions, to fathom the human condition. Perhaps the last
great, but no less failed, attempt at synthesis from the sociological
side was the "general theory of action" that Talcott Parsons tried
to forge while head of Harvard's interdisciplinary Social Relations

Department in the 1950s (Heims 1991). It is noteworthy that this project was advanced no more than a decade after the consolidation of the neo-Darwinian synthesis. I say this because the shadow cast over Darwinism by Nazi Germany's open endorsement meant that after World War II biology could have just as easily gone against as for Darwin (Weikart 2005). This turning away from Darwin re-opened Auguste Comte's prospect of biology becoming subsumed under sociology. A glimpse of that alternative future can be caught in the writings of the ubiquitous Julian Huxley.

Despite having coined the phrase "evolutionary synthesis," Huxley is nowadays treated as *persona non grata* amongst neo-Darwinists, not only because he continued to endorse social engineering even after Soviet excesses rendered the idea politically incorrect, but more importantly he saw eugenics as charting a neo-Lamarckian path to a convergence of human and divine powers (Provine 1988). Huxley regarded prior failures to turn biological knowledge to collective human advantage as products of misinformation, misunderstanding, impatience, and prejudice – but not an insuperable obstacle to human advancement. Indeed, Huxley regarded Darwin as having had more an eye to the past than to the future, which was unsurprising, considering his failure to understand the mechanisms of heredity.

Here we see then the roots of a secular sense of "intelligent design" that persists in science policy culture today. We saw in chapter 2 that Huxley, as UNESCO's first scientific director, advanced what he called a "transhumanist" vision that aimed to address our persistent economic and cultural problems by stretching the species' evolutionary limits. A half-century later, the US National Science Foundation updated this ambition in a long-term strategy to channel emerging developments in nanotechnology, biotechnology, information technology, and the cognitive sciences ("NBIC") for purposes of "enhancing human performance" (Roco and Bainbridge 2002). Versions of the strategy have been quickly adopted by the European Commission and the Japanese Science Foundations. Behind the enthusiasm for these initiatives is the prospect of a high-tech reinvention of such classic welfare state functions as healthcare and education through prosthetic innovations, aka "cyborgization."

Three large issues that go beyond even Huxley's original horizons will need to be addressed before this transhumanist reinvention of a global welfare state can be realized in a normatively desirable fashion:

1. *The Disabled*: Blindness, deafness, and other physical disabilities have inspired innovations which allow the disabled to interact effectively with full-bodied people, which in turn have expanded our conception of a common humanity (Fuller 2006b: ch. 10). However, there is increasing moral pressure toward the early identification and antenatal termination of disabilities. These measures are urged in the name of pre-empting "suffering," but whose exactly: the potentially disabled or the full-bodied people in whose presence they would live? Will this technological fix merely stunt the humanistic imagination to cope with only those whose lives are regarded as most convenient?

2. *The Aged*: Social stratification is strongly correlated with age, in terms of when one enters and exits a given role. This establishes the general contours of an individual's life-plan and provides the basis for patterns of mutual recognition that maintain stability through changes in societal personnel. But will a new basis for cross-generational social conflict emerge as, on the one hand, the age for traditionally adult activities is lowered and, on the other, the age for retirement from work, and life more generally, is indefinitely postponed?

3. *The Dead*: Society presupposes that the current generation eventually (literally or symbolically) dies, which provides an opportunity for a fresh perspective, given the difference between the younger and older generations' memories. Indeed, innovation often results from a confidence born of ignorance of the past. Moreover, at least in monotheistic cultures, the finality of death has focused the minds of individuals to do their best in the limited time they have. But will the attenuation of death – perhaps through some cyborgian transformation or reincarnation – remove this sense of progress and incentive for achievement?

The three groups of challenges imply the need for sociological replacements for the biological limits that an intelligently designed post-Darwinian future would aim to blur, if not entirely overcome. However, at the same time the most socially divisive features of Darwinism are unlikely to disappear without a fight, to which I finally turn.

A telling sign of the political struggle ahead to realize a post-Darwinian vision is the recent steady rise in the popularity of the United Kingdom's native fascist movement, the British National Party (BNP). We have seen that the UK originated and developed a strong biologistic perspective to social life without becoming racist in the process (Fuller 2006b: ch. 5). The line of descent from Darwin and Galton to the Webbs and Beveridge – that is, from evolution and eugenics to Fabian socialism and the welfare state – pretty much tracks the intellectual life of social scientists in metropolitan London from 1860 to 1960. Nevertheless, the result was *not* fascism, even though a home-grown fascist party received considerable press coverage during the Great Depression of the 1930s under the leadership of Oswald Mosley, a disenchanted upper-class Labour Party politician. The BNP picks up where Mosley left off. Its core constituency is still the white working class. The twist is that they are nowadays also the targets of widening participation schemes in higher education. The "skinheads" and "yobs" are effectively evolving. Little surprise, then, that the current, and most politically successful, head of the BNP, Nick Griffin, is a bespoke first-generation Cambridge graduate.

Nevertheless, the white working class are also the group least likely to benefit and most likely to suffer from the UK's global economic performance. A week after UK Employment Minister Margaret Hodge revealed the popularity of the BNP in her poor East London constituency in 2006, a YouGov survey found that most Britons supported BNP policies, with a mere 17 percent drop-off once respondents were told that the BNP espoused them. The only BNP policies they found truly abhorrent were the prioritized treatment of whites in welfare matters and the resettlement of recent immigrants in their countries of origin. But otherwise the respondents supported a political economy for Britain, insulating it from events elsewhere in the world.

This result is striking, given that the UK's three major parties – Labour, Tories, and the Liberal Democrats – nowadays stress their openness to the consequences of globalized liberalism: the free transit of people, products, and capital. Moreover, while recognizing the need for universal healthcare, education, and domestic security, all three parties have been reluctant to call for additional public revenue or to discriminate amongst its beneficiaries. The BNP would do both. It is truly a "nationalist" and a "socialist" party. This perhaps explains why Norman Tebbit, the Tory Party chair under Margaret Thatcher, could not see anything "right wing" in the BNP's 2005 campaign manifesto, a document that is usefully read alongside the Nazi Party's 1920 manifesto. The latter reflected the frustration of a nation essentially under house arrest by other nations – admittedly because of its instigation of World War I. Before we proceed, a proviso is in order: To make the most of such temporally specific comparisons, the reader should not be unduly influenced by the later horrible history of the Holocaust that was instigated once the Nazis came to power (cf. Fuller 2006b: ch. 14).

Read together, the Nazi and BNP manifestos share two common themes, one a spin on Marx, the other on Darwin. First, those who have worked to build up the nation are entitled to reap the lion's share of its collective benefits. Second, the state cannot maintain social order if it does not have a clear sense of who is and is not fit to live within its borders. Reflecting the tenor of our times, the BNP statement stresses Darwinist over Marxist themes. But then so did the Nazis', once the German economy worsened and the Nazis' own political situation improved. This common reliance on Darwinism serves as an uncomfortable reminder of an important source of the welfare state, namely, a geographical sense of biological enclosure.

To be sure, the history of the welfare state is usually told from the standpoint of the "Iron Chancellor" Otto von Bismarck, whose establishment of the first social security system in the 1880s was intended to immunize the heterogeneous Second German Reich against the threat of revolutionary socialism. Nevertheless, most welfare states were originally motivated by Darwinian concerns for the maintenance of biologically homogeneous populations, as arguably existed in, say, Scandinavia at the turn of the last century. As we have already seen, this sensibility was also strong in Patrick Geddes, the man who was nearly appointed to the UK's first sociology chair. Geddes abided by the Darwinian line that two species arise from a common population that have been allowed to reproduce in long-term segregation from each other. From this standpoint, a race is a proto-species that arguably will flourish best – in that sense, "evolve" into a proper species – if left segregated.

A comparable benchmark in US social science is the career of Edward A. Ross, the founder of the University of Wisconsin's Sociology Department, arguably the country's most accomplished. Ross conducted a half-century "Progressive" campaign under the approving eye of Theodore Roosevelt to organize labor, downsize capital, segregate races, limit immigration, and sterilize the incapacitated (Ross 1991: ch. 7). The racist and eugenicist sides of the US Progressive movement only started to be challenged once Blacks and Whites fought side by side in World War I, albeit in separate divisions. This was the closest America had yet gotten to a Bismarckian pretext for a welfare state, whereby the state owed a decent existence to everyone, regardless of social or ethnic background, who might be called to risk their lives in common defense of the national interest. It took another few global wars – culminating in Vietnam – for Americans to fully seal the Bismarckian pact with the passage of the Civil Rights Act of 1964.

The BNP manifesto is charitably read as saying that every nation should be encouraged to establish a welfare state within its own borders, but that the UK's welfare provision should not be abused by émigrés from nations that have failed to provide for their own. This would certainly explain the BNP's call to double the UK's foreign aid budget, including a generous resettlement scheme for recent immigrants. These costs would be outweighed by the prospective benefits, not least ecological sustainability, linguistic preservation, and global multiculturalism. But perhaps the BNP's most sophisticated welfare-based appeal turns on the likely impact of a "genetically diverse" (aka racially mixed) society on provision for the National Health Service. Here the BNP envisages either massive tax increases to cover a host of historically "non-British" ailments or, more likely in these liberal times, the devolution of health care to private providers. This would in turn disadvantage the native working-class population, most of whom are white.

Those (including myself) who look with sorrow at the passing of the welfare state need to confront the fact that major aspects of its legacy live on in the BNP. Indeed, the BNP appeals to much of the welfare state's original constituency that remains alienated by the mad political rush to the liberal center. It is worth recalling that William Beveridge was doing more than conferring a spurious sense of rigor when he called the discipline underwriting the British welfare state "social biology." Equally, more than a vote-getting ploy was behind the inclusion of "socialist" in the Nazi Party's name. The give-and-take of democratic elections, which foreground questions of who pays and who benefits, have always pushed socialist politics towards "biologizing" people, i.e. focusing on actual human populations rather than on an abstract sense of humanity. The BNP may continue to trade on racism and xenophobia, but it also plays to an important strand of the democratic socialist tradition, which we would be foolish to ignore in the coming years as the boundary dividing "the social" from "the biological" breaks down, courtesy of the neo-Darwinian synthesis.

Conclusion: The Larger Lessons

The historic conflict between science and religion has been more social than cognitive. Both science and religion have traditions of consensus and dissent, dogma and heresy – indeed, "normal" and "revolutionary" phases, in Kuhn's (1970) words (Fuller 2003: chs 10–11). The self-styled neo-Lamarckian philosopher of "creative evolution" Henri Bergson (1935) developed an entire cyclical theory of the moral order around this point, on the basis of which Karl Popper (1945) launched his famous polemical distinction between "closed" and "open" societies. For Bergson, what distinguishes a society as "closed" or "open" is its attitude toward a truth that transcends the sum of all experience. In other words: do members of the society identify more with modes of knowing that embed humans in an animal world responsive mainly to what lies before one's senses or with modes promising access to a more comprehensive view of reality closer to God's? In chapter 2, I spoke of this essential tension in the definition of humanity as our "carbon–silicon" duality.

After Kant, this choice between the closed and open visions has been canonically represented in epistemological terms as "empiricism" versus "rationalism," an intentionally secular distinction that obscures its religious implications. After all, the "rationalists" believed that the faculty of reason, especially our ability to intuit unity and order in reality, raised us above the animals, putting us closer to God. In contrast, the empiricists were profoundly skeptical of the existence of any faculties that pretended to knowledge beyond our habitual sensory encounters with the physical environment. As Kant realized, only the rationalists provided an incentive to do science in Newton's grand universal sense, yet the incentive fell short of proof – as the empiricists correctly stressed.

And 100 years later, reflecting the institutionalization of the special sciences that had occurred in the interim, the founder of the phenomenological movement, Edmund Husserl, partly recovered the religious roots of Kant's distinction by contrasting "immanent" and "transcendent" forms of objectivity as offering, respectively, immediate and mediated access to reality. Husserl's spin was aimed at what he saw as science's increasing arrogation of the divine standpoint over the 19th century through a discipline-based rationalized empiricism that overruled the testimony of personal experience. In effect, Husserl was worried that scientists had become secular iconophiles, having come to worship their theories rather than the reality that they were supposedly about. Husserl and others, notably his student Martin Heidegger, interpreted Germany's defeat in World War I as a world-historic rejection of such arrogant scientism. If 1859 marked science's opening salvo against religion's epistemic superiority in society, 1918 saw religion return fire.

However, a couple of generations after Husserl, Kant's distinction was re-spun in a more political direction in the hands of Popper's student, the social anthropologist Ernest Gellner (1989). Gellner did not challenge the empirical basis of knowledge as such, but the idea that truth could be automatically inferred from the most coherent set of locally validated empirical beliefs. Gellner drew on Popper's inspiration from John Stuart Mill to argue that a society's rationality can be gauged by its willingness to give a minority opinion a run for its money, since truth is unlikely to be closed under received wisdom. Here Gellner was actively resisting the emergence of scientific dogmatism in the West during the Cold War, defended by Popper's logical positivist colleagues, who became the analytic philosophy establishment in the US, distinguished anti-Communist scientists like Michael Polanyi, and, probably most influentially, the Harvard physicist-turned-historian Thomas Kuhn (Fuller 2000b). In Kuhn's hands, what the positivists defended as "foundations" and Polanyi quite openly as "dogma" and "commitment" became the protean but palatable term "paradigm."

An interesting feature of Gellner's analysis, which is relevant to the thesis of this book, is precisely the significance it accords *the book* as a vehicle of knowledge. The secularization and scientization of the Abrahamic faiths has been greatly facilitated by the spread of literacy and the demand that everyone read the Bible *for themselves*. It is too often forgotten that this is the context in which the "literal interpretation" of the Bible was originally promoted during the Protestant Reformation. After all, the most common way to read the Bible "literally" is as something that addresses the reader personally, which is to say, to live up to its message. This was indeed how, say, Newton read the Bible "literally," which impelled him to act as someone created "in the image and likeness of God" should, namely, by trying to comprehend the divine order (Harrison 1998).

Such a proactive understanding of the Bible, which equally informs evangelism, also held implications for the moral realm, since humans typically succumb to sin, when they fail to act in cases where their actions could have made a positive difference. Indeed, both major schools of philosophical ethics in the modern period, Kantianism and utilitarianism, were born of the human capacity, interpreted somewhat differently, to legislate for oneself as God does for reality as a whole (Schneewind 1997). The so-called *proactionary* principle associated with contemporary transhumanism, designed to counter the precautionary principle championed by the ecology movement, has recently redressed this late Enlightenment sensibility in futuristic garb (More 2005).

We often forget that in the century leading up to evolutionary ethics inspired by Darwin and Spencer, Kantianism and utilitarianism were seen as the ultimate Enlightenment legacy, the epitome of both the secularization and the scientization of Christian theology. The *summa* of this viewpoint was Henry Sidgwick's

Methods of Ethics (1874). However, evolution appeared to render ethics more reactive (to the physical environment) than proactive (in the name of ideals). Thus, Darwinists like Huxley and his progeny, who still believed that the proactive spirit of Kant and Mill was required to sustain the fruits of civilization, including science itself, have had to cast ethics in systematic opposition to natural selection. Nevertheless, this is an intellectually unsatisfying conclusion that seemingly pits "science" (as evolution) and "religion" (as ethics) in an endless Manichaean struggle, both in our brains and in society at large.

But, as we have seen in these pages, evolution's unresolved intellectual difficulties need to be balanced against the burden of proof that a religiously revived science of intelligent design would have to shift. Much of this burden is closely associated with the global hegemony of American culture – in this case, its legal and political culture: namely, the idea that science as a mode of inquiry is closed under a paradigm whose expert practitioners are capable of clearly demarcating authentic scientific claims from other pretenders to knowledge, especially religion. While this idea is understandable for reasons tied to the history of the United States, it remains unclear why such an idea should be imported uncritically elsewhere around the world by those seeking to resolve the demands of science and religion. After all, an excessive reliance on expert judgment would seem to cut against science's universal character as a form of knowledge, a source of "testability" as a criterion that scientific claims must meet. Unsurprisingly, the legal mind behind IDT, Phillip Johnson, has stressed – in a way that his fellow lawyer Francis Bacon would have appreciated – the need for standards for appraising the scientific status of knowledge claims that are not inherently biased against a newcomer.

At an intellectual level, IDT also faces substantial challenges. Foremost among them is that both friends and foes of the theory are profoundly ignorant of the centrality of intelligent design to the rise of modern science. There is much more to IDT than simply the sum of unsolved problems faced by modern evolutionary theory. Indeed, certain branches of the biological sciences, such as genetics, biochemistry, and the molecular biology, would have been difficult to motivate and sustain, had it not been for the fecundity of the design perspective for the conduct of scientific research. However, defenders of IDT have yet to reclaim these disciplines as belonging more properly to their camp than to a theory of life that stresses chance-based processes. This point is illustrated by the ease with which evolutionists continue to speak of nature as exhibiting "design without a designer," as if the literal meaning of this expression were self-evident.

In contrast, IDT defenders all too often presume the accuracy of how their opponents depict them, which leads them to focus exclusively on trying to falsify evolution by original research. Yet a biological science founded on intelligent design would radically reconfigure the disciplines. It would not simply be the flipside of the evolutionary paradigm. In particular, biology would be subsumed under a universal science of design as the discipline that pertains to a range of carbon-based (divinely created) technologies. There would be no need to claim a philosophical exception for the neo-Darwinian synthesis as a historical – or, as Popper would say, "historicist" – science, as evolution's defenders do today.

All of these issues are potentially translatable into high school science textbooks, where science versus religion is most hotly contested today. Here IDT is well placed to demonstrate how operating with religiously inspired, design-based assumptions has led to hypotheses whose empirical validity have been accepted even by those not sharing those assumptions. Newton and Mendel are obvious candidates for such didactic treatment. Driving home such a methodological lesson in science education is bound to be appreciated by the many religious students who are currently told that they must leave their beliefs at the door of the science classroom. However, at the moment, IDT instruction tends to be much more defensive, stressing that evolution does not falsify the claim that nature is the product of intelligent design, which invariably begs the question of whether IDT has any independent scientific merit of its own.

Finally, as stressed throughout this book, the historical character of the controversy between intelligent design and evolution matters for more than just pedagogical reasons. Our sense of what constitutes these two scientific positions, what distinguishes science from religion, and indeed Darwin's iconic status in Western culture are all affected by our understanding of the history of science. In the run-up to the bicentennial of Darwin's birth in 2009, the significance attached to the man behind the theory of evolution of natural selection becomes increasingly salient. Almost every prominent evolutionary theorist has now published a book interpreting Darwin for our times, including one that attempts to rewrite *Origin of Species* in Darwin's limpid style while updating his scientific references (Jones 1999). Worthy as these efforts have been, they nevertheless beg an important question: Were Darwin transported to today's world, and educated in such largely design-based sciences as genetics and molecular biology that were developed after his death, would he continue to interpret the balance of the evidence as telling against intelligent design in nature? Evolutionists take for granted that the answer would be "yes." However, if you believe (as I do) that the advent of genetics and molecular biology in the first half of the 20th century, culminating in the discovery of DNA's double-helix structure in 1953, outweighs the significance of Darwin's own work, you would be forced to conclude that Darwin would reinterpret natural selection as a design-based mechanism, possibly propelled by a divine engineer who could even command Newton's respect.

The first decade of the 21st century is an apt time to revisit the negotiated settlement between the geneticists and the natural historians that constitutes the neo-Darwinian synthesis. When Theodosius Dobzhansky invested nature with veto power over the hubris of scientifically fortified nation-states, "nature" amounted to a secular proxy for God that would hopefully stave off the sort of recklessness that had been responsible for two hot wars and one cold. However, both our scientific knowledge and our political exigencies have become more sophisticated and possibly more mature in the six or more decades since the neo-Darwinian synthesis was forged. It may be time to replace a diffuse appeal to natural selection which metaphorically shadows a divine presence with a humanly accountable sense of intelligent design, which implies that we take full responsibility for the planet – as if we were its creators.

Bibliography

Albury, R. (1993). "Ideas of Life and Death," in W. F. Bynum and R. Porter, eds., *Companion Encyclopedia of the History of Medicine*. London: Routledge, ch. 13.

Amin, S. (1991). "The Ancient World-System versus the Modern Capitalist World-System," *Review* 14: 349–86.

Armstrong, K. (2000). *The Battle for God: Fundamentalism in Judaism, Christianity and Islam*. New York: HarperCollins.

Balzer, W., Moulines, U., and Sneed, J. (1987). *An Architectonic for Science*. Dordrecht: Kluwer.

Barrow, J., and Tipler, F. (1986). *The Anthropic Cosmological Principle*. Oxford: Oxford University Press.

Basalla, G. (2006). *Civilized Life in the Universe*. Oxford: Oxford University Press.

Bateson, G. (1979). *Mind and Nature: A Necessary Unity*. New York: Bantam.

Becker, C. (1932). *The Heavenly City of the Eighteenth Century Philosophers*. New Haven: Yale University Press.

Behe, M. (1996). *Darwin's Black Box*. New York: Simon and Schuster.

Bergson, H. (1935). *The Two Sources of Morality and Religion*. London: Macmillan.

Bernal, J. D. (1939). *The Social Function of Science.* London: Macmillan.

Bhaskar, R. (2000). *From East to West: Odyssey of a Soul.* London: Routledge.

Blaut, J. M. (1993). *The Colonizer's Model of the World: Geographical Diffusion and Eurocentric History.* New York: Guilford Press.

Bosanquet, B. (1899). *The Philosophical Theory of the State.* London: Macmillan.

Bowler, P. (1988). *The Non-Darwinian Revolution: Reinterpreting a Historical Myth.* Baltimore: Johns Hopkins University Press.

Boyer, P. (2001). *Religion Explained: The Human Instincts that Fashion Gods, Spirits and Ancestors.* London: Heinemann.

Brandon, R., and Burian, R. eds. (1984). *Genes, Organisms, Populations: Controversies over the Units of Selection.* Cambridge, MA: MIT Press.

Brockman, J., ed. (2006a). *Intelligent Thought: Science versus the Intelligent Design Movement.* New York: Random House.

Brockman, J. (2006b). "Letter to Members of Congress," <http://www.edge.org/documents/congress_index.html>; posted on 22 June.

Brooke, J. H. (1991). *Science and Religion.* Cambridge, UK: Cambridge University Press.

Brown, A. (2006). "When Evolutionists Attack," *Guardian.* London: 6 March.

Campbell, D. T. (1988). *Methodology and Epistemology for the Social Sciences.* Chicago: University of Chicago Press.

Campbell, J. A. (1996). "John Stuart Mill, Charles Darwin, and the Culture Wars," *Intercollegiate Review* 31 (2): 44–51.

Carroll, S. (2005). *Endless Forms Most Beautiful: The New Science of Evo-Devo.* New York: Norton.

Carter, S. (1993). *The Culture of Unbelief.* New York: Doubleday.

Cassirer, E. (1950). *The Problem of Knowledge: Philosophy, Science, and History since Hegel.* New Haven: Yale University Press.

Cavalli-Sforza, L. (2000). *Genes, Peoples and Languages.* New York: Farrar, Straus and Giroux.

Ceccarelli, L. (2001). *Shaping Science with Rhetoric: The Cases of Dobzhansky, Schrödinger, and Wilson.* Chicago: University of Chicago Press.

Chisholm, R. (1982). "The Problem of the Criterion," in *The Foundations of Knowing.* Minneapolis: University of Minnesota Press.

Cohen, H. F. (1994). *The Scientific Revolution: An Historiographical Inquiry*. Chicago: University of Chicago Press.

Collini, S. (1976). "Hobhouse, Bosanquet and the State: Philosophical Idealism and Political Argument in England 1880–1918," *Past and Present* 72 (August): 86–111.

Collini, S. (2006). *Absent Minds: Intellectuals in Britain*. Oxford: Oxford University Press.

Collins, F. (2006). *The Language of God*. New York: Free Press.

Collins, R. (1998). *The Sociology of Philosophies*. Cambridge, MA: Harvard University Press.

Commager, H. S. (1978). *The Empire of Reason: How Europe Imagined and America Realized the Enlightenment*. London: Weidenfeld and Nicolson.

Coulter, A. (2006). *Godless: The Church of Liberalism*. New York: Random House.

Dahrendorf, R. (1995). *LSE: A History of the London School of Economics and Political Science 1895–1995*. Oxford: Oxford University Press.

Davies, P. (2006). *The Goldilocks Enigma: Why is the Universe Just Right for Life?* London: Allen Lane.

Davis, E. (1998). *TechGnosis: Myth, Magic and Mysticism in the Age of Information*. New York: Crown.

Davis, P., and Kenyon, D. (1993). *Of Pandas and People: The Central Questions of Biological Origins*, 2nd edn (orig. 1989). Richardson, TX: Foundation for Thought and Ethics.

Dawkins, R. (1976). *The Selfish Gene*. Oxford: Oxford University Press.

Dawkins, R. (1986). *The Blind Watchmaker*. Oxford: Oxford University Press.

Deane-Drummond, C., Szerszynski, B., and Grove-White, R., eds. (2003). *Re-Ordering Nature: Theology, Society and the New Genetics*. London: Continuum.

Dear, P. (2006). *The Intelligibility of Nature*. Chicago: University of Chicago Press.

Dembski, W. (1998). *The Design Inference*. Cambridge, UK: Cambridge University Press.

Dembski, W. (2001). *No Free Lunch: Why Specified Complexity Cannot Be Purchased Without Intelligence*. Lanham, MD: Rowman & Littlefield.

Dembski, W. (2004). "The Logical Underpinnings of Intelligent Design Theory," in Dembski and Ruse, eds. (2004), pp. 311–30.

Dembski, W., and Ruse, M., eds. (2004). *Debating Design: From Darwin to DNA*. Cambridge, UK: Cambridge University Press.

Dennett, D. (1995). *Darwin's Dangerous Idea: Evolution and the Meanings of Life*. New York: Simon & Schuster.

Dewey, J. (1953). *Reconstruction in Philosophy* (orig. 1920). Boston: Beacon Press.

Dickens, P. (2000). *Social Darwinism*. Milton Keynes, UK: Open University Press.

Dobzhansky, T. (1937). *Genetics and the Origin of Species*. New York: Columbia University Press.

Dobzhansky, T. (1967). *The Biology of Ultimate Concern*. London: Fontana.

Dobzhansky, T. (1973a). *Genetic Diversity and Human Equality*. New York: Basic Books.

Dobzhansky, T. (1973b). "Nothing in Biology Makes Sense Except in Light of Evolution," *The American Biology Teacher* 3 (March): 125–9.

Dorn, H. (1991). *The Geography of Science*. Baltimore: Johns Hopkins University Press.

Duhem, P. (1954). *The Aim and Structure of Physical Theory* (orig. 1914). Princeton: Princeton University Press.

Evans, G. (2003). *A Brief History of Heresy*. Oxford: Blackwell.

Fara, P. (2002). *Newton: The Making of Genius*. London: Macmillan.

Febvre, L. (1983). *The Problem of Unbelief in the Sixteenth Century* (orig. 1942). Cambridge, MA: Harvard University Press.

Fitelson, B., Stephens, C., and Sober, E. (1999). "How not to detect design," *Philosophy of Science* 66: 472–88.

Fleischacker, S. (2004). *A Short History of Distributive Justice*. Cambridge, MA: Harvard University Press.

Fodor, J. (1996). "Deconstructing Dennett's Darwin," *Mind and Language* 11 (3): 246–62.

Forrest, B., and Gross, P. (2004). *Creationism's Trojan Horse: The Wedge of Intelligent Design*. Oxford: Oxford University Press.

Foucault, M. (1979). *Discipline and Punish*. New York: Random House.

Frank, A. G. (1998). *Re-Orient: The Global Economy in the Asian Age*. Berkeley, CA: University of California Press.

Franklin, J. (2001). *The Science of Conjecture*. Baltimore: Johns Hopkins University Press.

Fuller, S. (1985). "Bounded Rationality in Law and Science," unpublished Ph.D. dissertation. University of Pittsburgh.

Fuller, S. (1988). *Social Epistemology*. Bloomington: Indiana University Press.

Fuller, S. (1992). "Epistemology Radically Naturalized: Recovering the Normative, the Experimental, and the Social," *Cognitive Models of Science*, ed. R. Giere. Minneapolis: University of Minnesota Press, pp. 427–59.

Fuller, S. (1993). *Philosophy of Science and Its Discontents*, 2nd edn (orig. 1989). New York: Guilford Press.

Fuller, S. (1997). *Science*. Milton Keynes: Open University Press.

Fuller, S. (1998a). "A Social Epistemology of the Structure-Agency Craze: From Content to Context," in A. Sica (ed.), *What Is Social Theory? The Philosophical Debates*. Oxford: Blackwell, pp. 92–117.

Fuller, S. (1998b). "An Intelligent Person's Guide to Intelligent Design Theory," *Rhetoric and Public Affairs* 1: 603–10.

Fuller, S. (2000a). *The Governance of Science: Ideology and the Future of the Open Society*. Milton Keynes: Open University Press.

Fuller, S. (2000b). *Thomas Kuhn: A Philosophical History for Our Times*. Chicago: University of Chicago Press.

Fuller, S. (2002). *Knowledge Management Foundations*. Woburn, MA: Butterworth-Heinemann.

Fuller, S. (2003). *Kuhn vs. Popper: The Struggle for the Soul of Science*. Cambridge, UK: Icon Books.

Fuller, S. (2004). "Descriptive vs Revisionary Social Epistemology: the Former as Seen by the Latter," *Episteme* 1/1: 23–34.

Fuller, S. (2005). "Another Sense of the Information Age," *Information, Communication and Society* 8: 459–63.

Fuller, S. (2006a). *The Philosophy of Science and Technology Studies*. London: Routledge.

Fuller, S. (2006b). *The New Sociological Imagination*. London: Sage.

Fuller, S. (2007). *New Frontiers in Science and Technology Studies*. Cambridge, UK: Polity.

Fuller, S., and Collier, J. (2004). *Philosophy, Rhetoric and the End of Knowledge: A New Beginning for Science and Technology Studies*, 2nd edn (orig. 1993, authored by Fuller). Mahwah, NJ: Lawrence Erlbaum Associates.

Funkenstein, A. (1986). *Theology and the Scientific Imagination*. Cambridge, UK: Cambridge University Press.

Galison, P. (1990). "Aufbau/Bauhaus: Logical Positivism and Architectural Modernism," *Critical Inquiry* 16: 709–52.

Geddes, P., and Branford, V. (1919). *The Coming Polity*. London: Williams and Norgate.

Gellner, E. (1989). *Plough, Sword and Book*. Chicago: University of Chicago Press.

Gilbert, W. (1991). "Towards a Paradigm Shift in Biology," *Nature* (10 January): 349: 99.

Gilder, G. (1981). *Wealth and Poverty*. New York: Bantam.

Gilder, G. (1989). *Microcosm: The Quantum Revolution in Economics and Technology*. New York: Simon and Schuster.

Glover, J. (1984), *What Sort of People Should There Be?* Harmondsworth, UK: Penguin.

Golan, T. (2004). *Laws of Men and Laws of Nature: The History of Scientific Expert Testimony in England and America*. Cambridge, MA: Harvard University Press.

Gould, S. J. (1989). *Wonderful Life*. New York: Norton.

Graff, G. (1992). *Beyond the Culture Wars*. New York: Norton.

Hacking, I. (1975). *The Emergence of Probability*. Cambridge, UK: Cambridge University Press.

Hacohen, M. (2000). *Karl Popper: The Formative Years 1902–1945*. Cambridge: Cambridge University Press.

Hahn, R. (2001). *Anaximander and the Architects*. Albany, NY: SUNY Press.

Halliday, R. J. (1968). "The Sociological Movement, the Sociological Society and the Growth of Academic Sociology," *Sociological Review* 16: 377–98.

Harris, M. (1968). *The Rise of Anthropological Theory*. New York: Thomas Crowell.

Harrison, P. (1998). *The Bible, Protestantism and the Rise of Natural Science*. Cambridge, UK: Cambridge University Press.

Hawking, S. (1988). *A Brief History of Time*. New York: Bantam.

Heims, S. (1991). *Constructing a Social Science for Postwar America*. Cambridge, MA: MIT Press.

Herrnstein, R., and Murray, C. (1994). *The Bell Curve: Intelligence and Class Structure in American Life*. New York: Free Press.

Hofstadter, R. (1955). *Social Darwinism in American Thought*. Boston: Beacon Press.

Hofstadter, R. (1962). *Anti-Intellectualism in American Life.* New York: Random House.

Hofstadter, R. (1965). *The Paranoid Style in American Politics.* New York: Alfred Knopf.

Horgan, J. (1996). *The End of Science: Facing the Limits of Knowledge in the Twilight of the Scientific Age.* Reading, MA: Addison Wesley.

Huff, T. (1993). *The Rise of Early Modern Science: Islam, China, and the West.* Cambridge, UK: Cambridge. University Press.

Huxley, J. (1947). *UNESCO: Its Purpose and Its Philosophy.* Washington: Public Affairs Press.

Israel, J. (2001). *Radical Enlightenment: Philosophy and the Making of Modernity 1650–1750.* Oxford: Oxford University Press.

Janik, A., and Toulmin, S. (1973). *Wittgenstein's Vienna.* New York: Simon & Schuster.

Johnson, P. (1991). *Darwin on Trial.* Chicago: Regnery Press.

Johnson, P. (2000). *The Wedge of Truth: Splitting the Foundations of Naturalism.* Chicago: Regnery Press.

Jones, S. (1999). *Almost Like a Whale: The Origin of Species Updated.* London: Doubleday.

Kauffman, S. (1995). *At Home in the Universe: The Search for the Laws of Self-Organization and Complexity.* Oxford: Oxford University Press.

Kay, L. (1993). *The Molecular Vision of Life. Caltech, the Rockefeller Foundation and the Rise of the New Biology.* Oxford: Oxford University Press.

Kay, L. (2000). *Who Wrote the Book of Life?* Cambridge, MA: MIT Press.

Kazin, M. (2006). *A Godly Hero: The Life of William Jennings Bryan.* New York: Alfred Knopf.

Kent, R. (1981). *A History of British Empirical Sociology.* Aldershot: Gower.

King, D. (1999). *In the Name of Liberalism.* Oxford: Oxford University Press.

Kitcher, P. (1978). "Theories, Theorists and Theoretical Change," *Philosophical Review* 87: 519–47.

Kitcher, P. (2001). *Science, Truth, and Democracy.* Oxford: Oxford University Press.

Knight, D. (2004). *Science and Spirituality: The Volatile Connection.* London: Routledge.

Koertge, N., ed. (1998). *House Built on Sand: Exposing Postmodern Myths about Science.* Oxford: Oxford University Press.

Koertge, N., ed. (2005). *Scientific Values and Civic Virtues.* Oxford: Oxford University Press.

Krimbas, C. (2001). "In Defence of Neo-Darwinism: Popper's 'Darwinism as a Metaphysical Program' Revisited," in R. Singh et al., eds., *Thinking about Evolution: Historical, Philosophical, and Political Perspectives.* Cambridge, UK: Cambridge University Press, pp. 292–308.

Kuhn, T. S. (1970). *The Structure of Scientific Revolutions,* 2nd edn (orig. 1962). Chicago: University of Chicago Press.

Kurzweil, R. (1999). *The Age of Spiritual Machines.* New York: Random House.

La Follette, M., ed. (1983). *Creationism, Science and the Law.* Cambridge, MA: MIT Press.

Lakatos, I. (1981). "'History of Science and Its Rational Reconstructions," in I. Hacking, *Scientific Revolutions.* Oxford: Oxford University Press, 1981, pp. 107–27.

Latour, B. (1988a). *The Pasteurization of France.* Cambridge, MA: Harvard University Press.

Latour, B. (1988b). "The Politics of Explanation," in S. Woolgar, ed., *Knowledge and Reflexivity.* London: Sage, pp. 155–76.

Laudan, L. (1981). *Science and Hypothesis.* Dordrecht: Kluwer.

Laudan, L. (1982). "Science at the Bar: Causes for Concern," *Science, Technology and Human Values* 7: 16–19.

La Vergata, A. (2000). "Biology and Sociology of Fertility, Reactions to the Malthusian Threat 1798–1933," *Clio Medica,* vol. 34; *Malthus, Medicine and Morality,* ed. B. Dolan, Amsterdam: Rodopi, pp. 189–222.

Lenski, R., Ofria, C., Pennock, R., and Adami, C. (2003). "The Evolutionary Origin of Complex Features," *Nature* 423: 139–44.

Lepenies, W. (1988). *Between Literature and Science: The Rise of Sociology.* Cambridge, UK: Cambridge University Press.

Lewontin, R. (1997). "The Demon Haunted World," *New York Review of Books* (9 January).

Livingstone, D. (1984). *Darwin's Forgotten Defenders: The Encounter between Evangelical Theology and Evolutionary Thought.* Grand Rapids: Eerdmans.

Löwith, K. (1949). *Meaning in History.* Chicago: University of Chicago Press.

Lynch, W. T. (2001). *Solomon's Child: Method in the Early Royal Society of London*. Palo Alto: Stanford University Press.

Lyotard, J.-F. (1983). *The Postmodern Condition* (orig. 1979). Minneapolis: University of Minnesota Press.

Masuzawa, T. (2005). *The Invention of World Religions*. Chicago: University of Chicago Press.

Mayr, E. (1942). *Systematics and the Origin of Species*. New York: Columbia University Press.

Menuge, A. (2004). *Agents under Fire: Materialism and the Rationality of Science*. Lanham, MD: Rowman and Littlefield.

Merton, R. K. (1977). *The Sociology of Science*. Chicago: University of Chicago Press.

Merz, J. T. (1965). *A History of European Thought in the 19th Century*, 4 vols (orig. 1896–1914). New York: Dover.

Milbank, J. (1990). *Theology and Social Theory*. Oxford: Blackwell.

Miller, K. (1999). *Finding Darwin's God*. New York: HarperCollins.

Mirowski, P. (2002). *Machine Dreams: Economics Becomes a Cyborg Science*. Cambridge, UK: Cambridge University Press.

Mooney, C. (2005). *The Republican War on Science*. New York: Simon & Schuster.

More, M. (2005). "The Proactionary Principle," <http://www.maxmore.com/proactionary.htm>.

Morris, S. C. (2003). *Life's Solution: Inevitable Humans in a Lonely Universe*. Cambridge, UK: Cambridge University Press.

Moss, L. (2003). *What Genes Can't Do*. Cambridge, MA: MIT Press.

Mumford, L. (1934). *Technics and Civilization*. New York: Harcourt, Brace & World.

National Academy of Sciences [NAS] (1999). *Science and Creationism: A View from the National Academy of Sciences*, 2nd edn. Washington, DC: NAS Press.

Nelkin, D. (1982). *The Creation Controversy: Science or Scripture in the Schools*. New York: Norton.

Noble, D. (1997). *The Religion of Technology: The Divinity of Man and the Spirit of Invention*. New York: Alfred Knopf.

Noelle-Neumann, E. (1981). *The Spiral of Silence*. Chicago: University of Chicago.

Olby, R. C. (1997). "Mendel, Mendelism and Genetics," *MendelWeb*, <http://www.mendelweb.org/MWolby.html>.

Oldroyd, D. R. (1980). *Darwinian Impacts.* Milton Keynes, UK: Open University Press.

Overton, W. (1983). "Opinion, *McLean vs Arkansas,*" in La Follette (1983), pp. 45–73.

Parsons, K. (2005). "Defending the Radical Center," in Koertge (2005), pp. 159–71.

Passmore, J. (1970). *The Perfectibility of Man.* London: Duckworth.

Paul, D. (1998). *The Politics of Heredity.* Albany, NY: SUNY Press.

Pickering, M. (1993). *Auguste Comte: An Intellectual Biography,* vol. 1. Cambridge, UK: Cambridge University Press.

Pinker, S. (2002). *The Blank Slate: The Modern Denial of Human Nature.* New York: Vintage.

Popper, K. (1945). *The Open Society and Its Enemies,* 2 vols. London: Routledge & Kegan Paul.

Popper, K. (1957). *The Poverty of Historicism.* London: Routledge & Kegan Paul.

Popper, K. (1972). *Objective Knowledge.* Oxford: Oxford University Press.

Porter, T. (1986). *The Rise of Statistical Thinking: 1820–1900.* Princeton: Princeton University Press.

Proctor, R. (1988). *Racial Hygiene: Medicine under the Nazis.* Cambridge, MA: Harvard University Press.

Provine, W. (1988). "Progress in Evolution and Meaning in Life," in M. Nitecki (ed.), *Evolutionary Progress.* Chicago: University of Chicago Press, pp. 49–72.

Pyenson, L., and Sheets-Pyenson, S. (1999). *Servants of Nature.* New York: Norton.

Rasmussen, N. (1994). "Surveying Evolution," *Metascience* 5: 55–60.

Ratzsch, D. (2001). *Nature, Design and Science.* Albany, NY: SUNY Press.

Reisch, G. (2005). *How the Cold War Transformed the Philosophy of Science.* Cambridge, UK: Cambridge University Press.

Richards, J., ed. (2002). *Are We Spiritual Machines?* Seattle: Discovery Institute.

Robertson, J. M. (1929). *A History of Freethought in the Nineteenth Century.* London: Watts & Co.

Roco, M., and Bainbridge, W. S., eds. (2002). *Converging Technologies for Enhancing Human Performance: Nanotechnology, Biotechnology, Informa-*

tion *Technology and Cognitive Science*. Arlington, VA: US National Science Foundation.

Roll-Hansen, N. (2005). *The Lysenko Effect: The Politics of Science*. Amherst, NY: Prometheus Books.

Rosen, R. (1999). *Essays on Life Itself*. New York: Columbia University Press.

Rosenberg, A. (2005). "Lessons from biology for philosophy of the human sciences," *Philosophy of the Social Sciences*. 35: 3–19.

Ross, D. (1991). *The Origins of American Social Science*. Cambridge, UK: Cambridge University Press.

Runciman, W. G. (1983–97). *A Treatise on Social Theory*, 3 vols. Cambridge, UK: Cambridge University Press.

Ruse, M. (1979). *The Darwinian Revolution: Science Red in Tooth and Claw*. Chicago: University of Chicago Press.

Ruse, M. (1983). "Creation-Science is Not Science," in LaFollette (1983), pp. 150–60.

Ruse, M. (1996). *Monad to Man: The Concept of Progress in Evolutionary Biology*. Cambridge, MA: Harvard University Press.

Ruse, M. (2003). *Darwin and Design: Does Evolution Have a Purpose?* Cambridge, MA: Harvard University Press.

Ruse, M. (2005). *The Evolution-Creation Struggle*. Cambridge, MA: Harvard University Press.

Sachs, J. (2005). *The End of Poverty: How We Can Make It Happen in Our Lifetime*. London: Penguin.

Schneewind, J. (1997). *The Invention of Autonomy: A History of Modern Moral Philosophy*. Cambridge, UK: Cambridge University Press.

Schroeder-Gudehus, B. (1990). "Nationalism and Internationalism," in R. C. Olby et al., eds., *Companion to the History of Modern Science*. London: Routledge, pp. 909–19.

Schubert, D. (2004). "Theodor Fritsch and the German (*völkische*) Version of the Garden City," *Planning Perspectives* 19/1: 3–35.

Scott, J., and Husbands, C. (2007). "Victor Branford and the Building of British Sociology," *Sociological Review* 55.

Secord, J. (2001). *Victorian Sensation: The Extraordinary Publication, Reception, and Secret Authorship of "Vestiges of the Natural History of Creation"*. Chicago: University of Chicago Press.

Segerstrale, U. (2000). *Defenders of the Truth*. Oxford: Oxford University Press.

Simon, H. (1977). *The Sciences of the Artificial.* Cambridge, MA: MIT Press.

Singer, P. (1975). *Animal Liberation.* New York: Random House.

Singer, P. (1999). *A Darwinian Left.* London: Weidenfeld & Nicolson.

Smith, J. M. (1998). *Shaping Life: Genes, Embryos and Evolution.* London: Weidenfeld & Nicolson.

Smocovitis, V. B. (1996). *Unifying Biology: The Evolutionary Synthesis and Evolutionary Biology.* Princeton: Princeton University Press.

Snyder, L. (2006). *Reforming Philosophy: A Victorian Debate on Science and Society.* Chicago: University of Chicago Press.

Sorokin, P. (1970). *Social and Cultural Dynamics* (orig. 1937–41). Boston: Porter Sargent.

Steele, D. R. (1986). *From Marx to Mises: Post-Capitalist Society and the Challenge of Economic Calculation.* La Salle, IL: Open Court.

Stegmüller, W. (1976). *The Structure and Dynamics of Scientific Theories.* Berlin: Springer-Verlag.

Sterelny, K. (2001). *Dawkins vs. Gould: Survival of the Fittest.* Cambridge, UK: Icon Books.

Strauss, L. (1952). *Persecution and the Art of Writing.* Chicago: University of Chicago Press.

Studholme, M. (2007). "Patrick Geddes: Father of Environmental Sociology," *Sociological Review* 55.

Teilhard de Chardin, P. (1955). *The Phenomenon of Man.* New York: Harper & Row.

Toulmin, S. (1972). *Human Understanding.* Oxford: Oxford University Press.

Toulmin, S. and Goodfield, J. (1965). *The Discovery of Time.* New York: Harper & Row.

Turbayne, C. (1962). *The Myth of Metaphor.* New Haven: Yale University Press.

Turner, S. P. (2003). *Liberal Democracy 3.0.* London: Sage.

Voegelin, E. (1968). *Science, Politics, and Gnosticism.* Chicago: Regnery Publishing.

Webster, C. (1975). *The Great Instauration: Science, Medicine and Reform, 1620–1660.* London: Duckworth.

Weikart, R. (2005). *From Darwin to Hitler.* New York: Macmillan.

Weinberg, S. (2001). *Facing Up: Science and Its Cultural Adversaries.* Cambridge, MA: Harvard University Press.

Werskey, G. (1988). *The Visible College: Scientists and Socialists in the 1930s*, 2nd edn (orig. 1978). London: Free Association Books.

White, M. (1957). *Social Thought in America: The Revolt against Formalism*. Boston: Beacon Press.

Wilson, E. O. (1975). *Sociobiology: The New Synthesis*. Cambridge, MA: Harvard University Press.

Wilson, E. O. (1998). *Consilience: The Unity of Knowledge*. New York: Alfred Knopf.

Wilson, E. O. (2006). *The Creation: An Appeal to Save Life on Earth*. New York: Norton.

Wolfson, H. (1976). *The Philosophy of the Kalam*. Cambridge, MA: Harvard University Press.

Wood, R., and Orel, V. (2005). "Scientific Breeding in Central Europe in the Early 19th Century: Background to Mendel's Later Work," *Journal of the History of Biology* 38: 239–72.

Woodward, T. (2003). *Doubts about Darwin: A History of Intelligent Design*. Grand Rapids: Baker Books.

Wright, R. (1999). "The Accidental Creationist: Why Stephen Jay Gould is Bad for Evolution," *The New Yorker* (13 December).

Young, R. (1985). *Darwin's Metaphor*. Cambridge, UK: Cambridge University Press.

Index